# Shale Oil, Tar Sands, and Related Fuel Sources

**Teh Fu Yen,** EDITOR

*University of Southern California*

A symposium co-sponsored by
the Division of Fuel Chemistry
and the Division of Petroleum
Chemistry, Inc. at the 167th
Meeting of the American Chemical
Society, Los Angeles, Calif.,
April 3–5, 1974

ADVANCES IN CHEMISTRY SERIES **151**

AMERICAN CHEMICAL SOCIETY

WASHINGTON, D. C.        1976

An

Library of Congress CIP Data

Shale oil, tar sands, and related fuel sources.
  (Advances in chemistry series; 151 ISSN 0065-2393)

  Includes bibliographical references and index.

  1. Shale oils—Congresses. 2. Oil-shales—Congresses.
3. Oil sands—Congresses. 4. Petroleum, Synthetic—Con-
gresses.
  I. Yen, Teh Fu, 1927-    . II. American Chemical
Society. Division of Fuel Chemistry. III. American
Chemical Society. Division of Petroleum Chemistry. IV.
Series:

QD1.A355 no. 151 [TP699]      540'.8s      [533'.282]
ISBN 0-8412-0282-6            ADCSAJ       151 1-183

# Advances in Chemistry Series

**Robert F. Gould,** *Editor*

# FOREWORD

ADVANCES IN CHEMISTRY SERIES was founded in 1949 by the American Chemical Society as an outlet for symposia and collections of data in special areas of topical interest that could not be accommodated in the Society's journals. It provides a medium for symposia that would otherwise be fragmented, their papers distributed among several journals or not published at all. Papers are refereed critically according to ACS editorial standards and receive the careful attention and processing characteristic of ACS publications. Papers published in ADVANCES IN CHEMISTRY SERIES are original contributions not published elsewhere in whole or major part and include reports of research as well as reviews since symposia may embrace both types of presentation.

# CONTENTS

# PREFACE

Approximately 100 years ago, during the construction of a railroad, a workman piled pieces of excavated rock to enclose a campfire and discovered that the stone ignited. This incident was recorded in the March 1874 issue of *Scientific American* as the first discovery of oil shale as an energy source in this continent.

About 50 years later, the American Chemical Society published the first book on "Shale Oil," as the ACS Monograph Series No. 25, by Ralph H. McKee of Columbia University. His remarks still hold true to this day, as evidenced by an excerpt from his preface, dated April 1925.

"The most concentrated and the most easily used fuels have been those most sought for and most used. The discovery of a new and more concentrated fuel has been followed each time by a notable increase in the number and importance of devices for the conversion of heat into other forms of energy valuable to industry and life."

"This volume has been written because petroleum has achieved such great importance as a concentrated fuel in American industrial life and because oil from oil shale promises in the immediate future to supplement the supplies of well petroleum, and, in the more distant future, when the supply of well petroleum gradually fails, to furnish the bulk of the oil needed."

Another 50 years have elapsed during which oil shale, tar sands, and other supplementary fuels have not been fully exploited, despite the fact that this country is facing an energy shortage. It was an opportune time when the ACS Division of Fuel Chemistry decided to sponsor a symposium on "Shale Oil, Tar Sands, and Related Fuel Sources" in April 1974 at Los Angeles. Many of the papers presented at that symposium have been updated for the present volume.

My heartiest thanks to Howard B. Jenson of ERDA for his generous assistance and my sincerest acknowledgments to all the contributors for their patience and cooperation. I would also like to mention the assistance of many of my students, associates, and especially the clerical work of Darlene Baxter, Donna Jue, and Barbara James.

Teh Fu Yen

University of Southern California
Los Angeles, Calif.
March 19, 1976

# Characteristics of Synthetic Crude from Crude Shale Oil Produced by *in Situ* Combustion Retorting

R. E. POULSON, C. M. FROST, and H. B. JENSEN

U.S. Energy Research and Development Administration, Laramie Energy Research Center, Laramie, Wyo. 82071

*A synthetic crude prepared by hydrogenating the naphtha, the light oil, and the heavy oil fractions obtained from in situ crude shale oil by distillation and coking of the 850°F+ residuum was characterized by examining the fractions. The heavy oil had 935 ppm nitrogen of which 40% was pyridine-type nitrogen, and 60% was pyrrole-type nitrogen. The light oil had 79 ppm nitrogen, all of which was pyridine-type nitrogen. The naphtha had less than 1 ppm nitrogen, which was not characterized. The saturate content of the fractions was—naphtha, 87%; light oil, 77%; and heavy oil, 73%. In addition, the heavy oil contained 6% olefins and 2% polar material.*

The nitrogen contents of *in situ* crude shale oils may be somewhat lower than those of crude shale oils produced in other retorts (1); however, these *in situ* oils still contain more than twice as much nitrogen as high nitrogen petroleum crude oils. Because existing refineries would not be able to cope with the high nitrogen content of shale oil if it were a substantial portion of the refinery feed, the National Petroleum Council (NPC) has suggested (2) that crude shale oil be upgraded at the retorting site by a catalytic hydrogenation process to produce a synthetic, premium feedstock called "syncrude." The production of such a syncrude from *in situ* crude shale, a description of its bulk properties, and a comparison of its properties to those of an NPC-type syncrude have been covered in Chapter 6 of this volume. This paper reports the compound-type characteristics of that syncrude produced by Frost (3) by catalytic hydrogenation of *in situ* crude oil. Special attention has been devoted to the nitrogen-compound types that are likely to be present in a syncrude because it is these compounds with which a refiner will have to deal if he uses this or a similar syncrude as his refinery feed.

In addition to reporting the nitrogen-compound types present in this syncrude, this paper also reports on the nitrogen types in intermediate hydrogenation products in order to relate this study to other studies (4, 5, 6, 7, 8) which have shown that the efficacy of nitrogen removal depends upon the nitrogen types in the charge stock. Earlier studies were on pure compounds, or on charge stocks spiked with pure compounds, or on nitrogen-containing stocks and were concerned with nitrogen removals approaching 80%. The syncrude described in the present work represents a case approaching 95–99% nitrogen removal.

## Experimental

**Preparation of the Samples.** The synthetic crude oil (syncrude) (3) used in this study was prepared by hydrogenating the naphtha (IBP–350°F), the light oil (350°–550°F), and the heavy oil (550°–850°F) fractions that had previously been obtained from *in situ* crude shale oil by distillation and coking of the vacuum residuum. The syncrude is the proportioned sum of these hydrogenated products. The heavy oil used in this study was the 550°F+ material from the heavy oil hydrogenation. The light oil was the 350°F+ material from the hydrogenation of the light oil from the distillation step combined with the 350°–550°F material from the heavy oil hydrogenation. The 175°–350°F heavy naphtha was the 175°F+ material from the hydrogenation of the combined IBP–350°F naphtha from the distillation and the heavy naphthas from both the heavy oil and the light oil hydrogenations. The C$_5$–175°F light naphtha was the material with that boiling range from each of the three hydrogenations.

In addition to using these four fractions in the syncrude characterization the nitrogen compounds in three intermediate hydrogenation fractions were characterized in order to relate this denitrification study to other such studies. These materials were the light oil from the heavy oil hydrogenation, the 175°–350°F heavy naphtha from the heavy oil hydrogenation, and the 175°–350°F heavy naphtha from the light oil hydrogenation.

The heavy oil, which contained nearly 90% of the nitrogen in the syncrude, was fractionated by liquid displacement chromatography on Florisil. The nonpolar, nonnitrogen-containing hydrocarbons were washed from the Florisil column with *n*-heptane, a very weak base concentrate was displaced with benzene, and a weak base concentrate was displaced with benzene–methanol azeotrope.

**Analytical Methods.** Total nitrogen values were determined with a reductive, hydrogen–nickel pyrolysis tube and an ammonia micro-coulometer. Nonaqueous potentiometric titration (*6, 9, 10, 11*) was used to classify the nitrogen compounds into weak base ($pK_a$ +2– +8), very weak base ($pKa$ −2–+2), and neutral types. Infrared spectrometry (*10, 11, 12*) was used to determine the concentration of pyrrolic nitrogen (nonhydrogen–bonded N-H). Colorimetry (*10, 13, 14*) was used to determine pyrroles and indoles with unsubstituted $\alpha$ or $\beta$ positions. These methods classified the nitrogen compounds into weak bases such as pyridines (including quinolines, 5,6,7,8-tetrahydroquinolines, and acridines) and as arylamines (including 1,2,3,4-tetrahydroquinolines, 2,3-dihydroindoles, and anilines); into very weak base pyrroles and indoles with an $\alpha$ or $\beta$ position unsubstituted; and into neutral carbazoles without N-substitution. Low voltage mass spectrometry and high resolution mass spectrometry allowed classification of the remaining nitrogen compounds into either pyrrole types with $\alpha$ and $\beta$ positions substituted or carbazoles with N-substitution.

Hydrocarbon types were estimated using the substractive method of Poulson (*15, 16*) for the fractions boiling above 175°F. The hydrocarbon compound composition of the $C_5$–175°F naphtha was determined by gas chromatography. Paraffin and naphthene contents of the 175°–350°F naphtha and of the 350°–550°F light oil were calculated from mass spectra. Liquid displacement chromatography on Florisil was used to determine the amount of polar material in the 550°–850°F heavy oil.

### Results and Discussion

**Hydrocarbon-type Characterization.** Table I lists the four fractions, their wt % of the syncrude, and their hydrocarbon-type compositions. The values for polar material for the two naphthas and the light oil are estimates based on their nitrogen contents. The polar material value for the heavy oil is based on the recovered weights from the Florisil separa-

### Table I. Hydrocarbon Types in Syncrude Fractions

| Boiling Range | Name | Wt % of Crude | Hydrocarbon Type (Wt % of Fraction) | | | | |
|---|---|---|---|---|---|---|---|
| | | | Par-affins | Naph-thenes | Ole-fins | Aro-matics | Polar Material |
| $C_5$–175°F | light naphtha | 3 | 71.8 | 20.5 | 0.0 | 7.7 | <0.001 |
| 175°–350°F | heavy naphtha | 21 | 42.8 | 43.4 | 0.0 | 13.8 | <0.001 |
| 350°–550°F | light oil | 49 | 51.5 | 25.0 | 0.0 | 23.5 | <0.01 |
| 555°–850°F | heavy oil | 27 | 72.7[a] | | 6.0 | 19.2 | 2.1 |

[a] Includes naphthenes.

**Table II.  Microcoulometric and Titration Data for Syncrude Fractions**

| Fraction | Total Nitrogen in Fraction (ppm) | Nitrogen Type (Wt % of Nitrogen in Fraction) | | | |
|---|---|---|---|---|---|
| | | Weak Base | Very Weak Base | Neutral | Arylamine |
| Light naphtha | <0.5 | — | — | — | — |
| Heavy naphtha | 0.8 | — | — | — | — |
| Light oil | 79 | 100 | 0 | 0 | 0 |
| Heavy oil | 935 | 40.1 | 13.9 | 46.0 | 0 |

tion. As shown in Table I, all fractions of the syncrude have appreciable amounts of aromatics after the hydrogenation even though the nitrogen has been largely removed. Only the heavy oil has a detectable concentration of olefinic hydrocarbons. A reference to this olefinic nature is made later in this paper.

**Nitrogen-type Characterization.** SYNCRUDE FRACTIONS. Table II lists the microcoulometric and titration data for the four syncrude fractions. The 79-ppm nitrogen in the light oil was shown to be all pyridine-type nitrogen because it did not acetylate when acetic anhydride was used as the titration solvent. No further characterization of this nitrogen was carried out. No acetylatable arylamines were found in this fraction or in the heavy oil, although Brown (7) found that anilines made up nearly one third of the tar-base concentrate from a recycle, hydrocracked shale oil naphtha (total nitrogen in the naphtha was approximately 1000 ppm). In addition Silver (5) reports that in denitrification of shale gas oil to about 80% removal of nitrogen, arylamines appear to build up relative to the other nitrogen types in the total product oil.

NITROGEN CONCENTRATES. Table III lists the recovery of the heavy oil nitrogen in the two Florisil concentrates and also the concentration of nitrogen types in each of these two concentrates. The weak base and very weak base types are determined by nonaqueous potentiometric

**Table III.  Nitrogen Distribution in Heavy Oil Concentrates Displaced from Florisil**

| Concentrate | Recovery (Wt % of Nitrogen in the Heavy Oil) | Nitrogen Type (Wt % of Nitrogen in Concentrate) | | | | |
|---|---|---|---|---|---|---|
| | | N–H | Pyrrolic | Weak Base | Very Weak Base | Neutral |
| Very weak base concentrate | 61.5 | 63.4 | 6.8 | 0.0 | 19.8 | 80.2 |
| Weak base concentrate | 38.4 | 2.3 | <0.01 | 97.4 | .0 | 2.6 |

titration, the neutral types by difference, the N-H types by infrared spectrometry, and the pyrrolic types by colorimetry. As shown in Table III, the concentrate-labeled very weak base has about one-fifth very weak base and four-fifths neutral nitrogen compounds, and the nitrogen in the concentrate-labeled weak base is nearly all weak base nitrogen with less than 3% being neutral nitrogen.

Very Weak Base Nitrogen Characterization. Table IV is a summary of the data obtained from low voltage mass spectrometry and from high resolution mass spectrometry on the very weak base concentrate from the Florisil separation. Because there are no titratable weak bases (Table III) the Z series ions as shown in Table IV can be labeled as very weak base or neutral nitrogen types. The names of the very weak base nitrogen compounds reflect only the degree of hydrogen deficiency necessary to yield the proper Z series. The requirements for the proper Z series could be met by having olefinic bonds in either the cycloalkano rings or in an alkyl substituent on the ring system. For example, the $Z = -13$ labeled as dicycloalkanoindoles could also be correctly labeled

### Table IV. Mass Spectral Data for Very Weak Base Concentrate

| Z Series[a] | Compound Type | Percent of Ionization[b] |
|---|---|---|
| −5 | cycloalkanopyrroles | <1 |
| −7 | dicycloalkanopyrroles | 6 |
| −9 | indoles | 7 |
| −11 | cycloalkanoindoles | 3 |
| −13 | dicycloalkanoindoles | 13 |
| −15 | carbazoles | 54 |
| −17 | cycloalkanocarbazoles | 17 |
| −19 | dicycloalkanocarbazoles | <1 |

[a] From high resolution spectrum.
[b] From low voltage spectrum.

as monocycloalkenoindoles. The inclusion of an olefinic bond in the molecule is reasonable when one considers that 6% of the heavy oil hydrocarbon molecules are olefinic (Table I).

Interpretation of these mass-spectral data in Table IV combined with the titration data from Table III allows additional inference concerning the characteristics of the nitrogen compounds present in the very weak base concentrate. The sum of the compound types listed as pyrroles and indoles ($Z = -7, -9, -11,$ and $-13$) amounts to 29% of the total. These compound types have given (11) very weak base titers of about 70% of theoretical; thus it appears likely that the 19.8% titer for very weak bases in Table III may come from the titration of the pyrroles and indoles in this concentrate. In addition, they are not N-substituted because pyrrole type nitrogen titrates as weak base nitrogen, and there is no weak base titer for this concentrate.

A further chracterization of these pyrrole type nitrogen compounds in the very weak base concentrate can be made by using the colorometric pyrrolic nitrogen value of 6.8% (Table III) as the value for $\alpha,\beta$-unsubstituted pyrrole type compounds. This leaves 22.2% of the nitrogen in pyrroles and indoles which have both $\alpha$- and $\beta$-substitution. The $\alpha,\beta$-unsubstituted pyrroles and indoles also have no N-substitution because these N-substituted compounds would titrate as weak bases and not as very weak bases. The lack of N-substitution on the pyrroles and indoles is consistent with the research of Jacobson (18, 19) who reported that N-alkylpyrroles and N-alkylindoles thermally and irreversibly isomerize to give the $\alpha$ and $\beta$ alkyl isomers and therefore would not likely be present in crude shale oil.

Table III shows that 63.4% of the nitrogen in the very weak base concentrate is N-H nitrogen. Of this 63.4% of the nitrogen, 29% has already been characterized as being in pyrroles and indoles without N-substitution. This leaves 34.4% of the nitrogen to be in carbazoles without N-substitution. Table IV shows that 71% of the nitrogen is in carbazole type compounds, and if 34.4% has no N-substitution then 36.6% has N-substitution. Table V is a summary of these findings. Thus

**Table V.  Summary of Nitrogen Types in the Very Weak Base Nitrogen Concentrate**

| Nitrogen Type Compounds | Percent of Fraction |
| --- | --- |
| Pyrroles or indoles with either $\alpha$ or $\beta$ positions unsubstituted; no N-substitutions | 6.8 |
| Pyrroles or indoles with substitutions in both $\alpha$ and $\beta$ positions; no N-substitutions | 22.2 |
| Carbazoles with no N-substitution | 34.4 |
| Carbazoles with N-substitution | 36.6 |

we see that one third of the weak bases is one- and two-aromatic-ring heterocyclics and two thirds is three-aromatic-ring heterocyclics.

**Weak Base Nitrogen Characterization.** Table VI summarizes the mass spectral data on the weak base concentrate. As was true for the very weak base fraction, the compound types reflect only the degree of hydrogen deficiency necessary to achieve the proper Z series, and as in Table IV, this hydrogen deficiency could be achieved by olefinic bonds in the molecule. We can see from Table III that nearly all (97.4%) of the nitrogen in this concentrate titrates as weak base nitrogen; hence, the compound types listed in Table VI are generally consistent with that titration. However, Table III does show that 2.3% of the nitrogen in this fraction exhibits an N-H character. Because there is no very weak

**Table VI. Mass Spectral Data for Weak Base Concentrate**

| Z Series[a] | Compound Type | Percent of Ionization[b] |
|---|---|---|
| −5 | pyridines | 35 |
| −7 | cycloalkanopyridines | 25 |
| −9 | dicycloalkanopyridines | 12 |
| −11 | quinolines | 13 |
| −13 | cycloalkanoquinolines | 8 |
| −15 | dicycloalkanoquinolines | 3 |
| −17 | acridines | 4 |
| −19 | cycloalkanoacridines | <1 |

[a] From high resolution spectrum.
[b] From low voltage spectrum.

base titer, it can be assumed that this N-H character is in carbazoles either in the Z = −15 or −17 series. This indicates that there was some tailing of the carbazoles into the weak base fraction. Contrasted to the ring structures in the very weak base fraction in which one-ring and two-ring structures accounted for only one third of the fraction and three-ring structures two thirds, the weak bases are composed of two-thirds one-aromatic-ring structures and one-third two-ring and three-ring structures. Dinneen (*20*) also showed this preponderance of one-ring materials in a shale oil gas oil as did Poulson (*10*) in a shale oil light distillate.

CHARACTERIZATION OF INTERMEDIATE FRACTIONS. Nonaqueous potentiometric titration was used to characterize the nitrogen compounds in three intermediate fractions from the production of the syncrude. Table VII lists these three fractions, their source, and the titration results. Also

**Table VII. Microcoulometer and Titration Data for Selected, Intermediate, and Syncrude Fractions**

| Fraction | Total Nitrogen in Fraction (ppm) | Nitrogen Type (Wt % of Nitrogen in Fraction) | | | |
|---|---|---|---|---|---|
| | | Weak Base | Very Weak Base | Neutral | Arylamine |
| *Heavy Oil Hydrogenation* | | | | | |
| Heavy oil | 935 | 40.1 | 13.9 | 46.0 | 0 |
| Light oil | 1220 | 78.7 | 11.5 | 9.8 | 12.3[a] |
| Heavy naphtha | 299 | 100.0 | 0 | 0 | 0 |
| *Light Oil Hydrogenation* | | | | | |
| Light oil | 79 | 100.0 | 0 | 0 | 0 |
| Heavy naphtha | 53 | 100.0 | 0 | 0 | 0 |

[a] Included in weak base nitrogen.

listed in this table are the syncrude fractions produced during the hydrogenation.

The first fraction listed is the 550°F+ heavy oil produced by hydrogenation of the 550°–850°F heavy oil from distillation and coking of the *in situ* crude oil. This is the same fraction listed in Tables I and II as the syncrude heavy oil fraction. The second fraction listed in Table VII is the 350°–550°F light oil produced in the foregoing hydrogenation, and the third fraction is the 175°–350°F heavy naphtha produced in the same hydrogenation.

The fourth fraction in Table VII is the 350°–550°F light oil produced from the hydrogenation of the 350°–550°F light oil resulting from the distillation and coking of the *in situ* crude to which had been added an aliquot amount of the light oil shown as the second fraction in Table VII. The fifth fraction is the 175°–350°F heavy naptha from this hydrogenation. Only the second, third, and fifth listed fractions are intermediate fractions; the first and fourth are final, syncrude ones.

The data in Table VII show three things. First, it is evident that when a shale oil stock is hydrocracked to lower boiling material, the nitrogen content of the lower boiling fraction is higher than when that same boiling range material is hydrogenated. For example, the light oil from the heavy oil hydrogenation has 1220 ppm nitrogen whereas the light oil from the light oil hydrogenation is only 79 ppm nitrogen. Second, the light oil from the light oil hydrogenation and both of the naphthas have only weak base nitrogen, and this is all pyridine type because none of these fractions contains acetylatable amines. And, third, the light oil from the heavy oil hydrogenation has arylamine type weak bases, neutral nitrogen compounds, and very weak bases as well as pyridine type weak bases. These data indicate that an unspecified but appreciable amount of hydrocracking can and does precede denitrification reactions. In general, arylamines are not found in appreciable concentrations in shale oil distillates but are presumed to be produced by hydrogenation of a five-membered nitrogen ring to give a 2,3-dihydroindole type arylamine and by hydrogenation of a six-membered nitrogen ring to give a 1,2,3,4-tetrahydroquinoline type arylamine (*4, 5, 6, 10*). Cracking of this saturated, nitrogen-containing ring with the nitrogen remaining attached to the aromatic moiety results in an aniline type arylamine. The presence of arylamines in the light oil fraction from the heavy oil hydrogenation is consistent with Silver's (*5*) finding them in the total product from shale gas oil hydrogenation at nitrogen removals approaching the 80% level. The fact that the arylamines are not in either of the napta materials shown in Table VII nor in the naphthas shown in Table II indicates that arylamines are converted to hydrocarbons and ammonia when nitrogen conversion approaches 95%.

*Summary*

The synthetic crude was produced by hydrogenating the IBP–350°F naphtha, the 350°–550°F light oil, and the 550°–850°F heavy oil fractions obtained from *in situ* crude shale oil by distillation followed by coking of the 850°F+ residuum. Characterization of the syncrude was accomplished by examining the following fractions: $C_5$-175°F light naphtha, 175°–350°F heavy naphtha, 350°–550°F light oil, and 550°–850°F heavy oil.

The light naphtha comprised 3% of the syncrude and contained 72% paraffins, 20% naphthenes, 8% aromatics, and less than 0.5 ppm nitrogen. The heavy naphtha comprised 21% of the syncrude and contained 43% paraffins, 43% naphthenes, 14% aromatics, and less than 1 ppm nitrogen. The light oil comprised 49% of the syncrude and contained 51% paraffins, 25% naphthenes, 24% aromatics, and 79 ppm nitrogen. The heavy oil comprised 27% of the syncrude and contained 73% saturates, 6% olefins, 19% aromatics, 2% polar compounds, and 935 ppm nitrogen.

The nitrogen compounds in the two naphthas were not characterized. The nitrogen in the light oil was shown to be pyridine type nitrogen with no detectable arylamine type nitrogen.

The nitrogen compounds in the heavy oil were shown to be 40% weak base, 14% very weak base, and 46% neutral compounds. Mass spectrometry was used to classify the weak bases as 72% pyridines (one aromatic ring), 24% quinolines (two aromatic rings), and 4% acridines (three aromatic rings). Mass spectrometry combined with infrared analysis and pyrrolic-nitrogen determination were used to classify the very weak base and neutral nitrogen compounds as 7% pyrroles or indoles with either or both of the $\alpha$ and $\beta$ positions open and not N-substituted, 22% pyrroles or indoles with substitution on both $\alpha$ and $\beta$ positions but no N-substitutions, 34% carbazoles with no N-substitutions, and 37% N-substituted carbazoles.

In addition to characterizing the four fractions of the final syncrude product, three intermediate fractions were also characterized. The heavy naphtha from the light oil hydrogenation and the heavy naphtha from the heavy oil hydrogenation both contained characterizable amounts of nitrogen compounds which were weak bases of the pyridine type. The light oil from the heavy oil hydrogenation had the greatest concentration of nitrogen of any of the fractions examined, 1,220 ppm. Nonaqueous titration showed these nitrogen compounds to be 66% weak base pyridines, 12% weak base anilines (arylamines), 12% very weak bases, and 10% neutral compounds. This material contained the only evidence of arylamines in the product from these hydrogenations, which represent removals of 95–99.9% of the nitrogen of the charge stock.

## Literature Cited

1. Jensen, H. B., Poulson, R. E., Cook, G. L., *Am. Chem. Soc., Div. of Fuel Chem., Preprints* **15** (1) 113 (1971).
2. U.S. Energy Outlook, "An Interim Report," National Petroleum Council (1972) **2**, 80.
3. Frost, C. M., Poulson, R. E., Jensen, H. B., ADV. CHEM. SER. (1976) **151**, 77.
4. Frost, C. M., Jensen, H. B., *Am. Chem. Soc., Div. of Petrol. Chem., Inc., Preprints* **18** (1) 119 (1973).
5. Silver, H. F., Wang, N. H., Jensen, H. B., Poulson, R. E., "Air Pollution Control and Clean Energy," *AIChE Symp. Ser.* (1975), in press.
6. Koros, R. M., Banic, S., Hoffman, J. E., Kay, M. I., *Am. Chem. Soc., Div. of Petrol. Chem., Inc., Preprints* **12** (4) B165 (1967).
7. Flinn, R. A., Larson, O. A., Beuther, H., *Pet. Refiner* (1963) **42**, 129.
8. Doelman, J., Vlugter, J. C., *Proc. World Pet. Cong., 6th, Section III, Paper 12* (1963) Frankfurt/Main, Germany.
9. Buell, B. E., *Anal. Chem.* (1967) **39**, 756.
10. Poulson, R. E., Jensen, H. B., Cook, G. L., *Am. Chem. Soc., Div. of Petrol. Chem., Inc., Preprints* **16** (1) A49 (1971).
11. Silver, H. F., Wang, N. H., Jensen, H. B., Poulson, R. E., *Am. Chem. Soc., Div. of Petrol. Chem., Inc., Preprints* **17** (4) G74 (1972).
12. Pozefsky, A., Kukin, I., *Anal. Chem.* (1955) **27**, 1466.
13. Thompson, R. B., Symon, T., Wankat, C., *Anal. Chem.* (1952) **24**, 1465.
14. Muhs, M. A., Weiss, F. T., *Anal. Chem.* (1958) **30**, 259.
15. Poulson, R. E., Jensen, H. B., Duvall, J. J., Harris, F. L., Morandi, J. R., *Anal. Instrum.* (1972) **10**, 193.
16. Jackson, L. P., Allbright, C. S., Jensen, H. B., *Anal. Chem.*, in press.
17. Brown, D., Earnshaw, D. G., McDonald, F. R., Jensen, H. B., *Anal. Chem.* (1970) **42**, 146.
18. Jacobson, Jr., I. A., Jensen, H. B., *U.S. Bur. Mines Rep. Invest.* (1966) **672**, 50.
19. Jacobson, Jr., I. A., *U.S. Bur. Mines Rep. Invest.* (1971) **7529**, 8.
20. Dinneen, G. U., Cook, G. L., Jensen, H. B., *Anal. Chem.* (1968) **40**, 1295.

RECEIVED December 16, 1974. The work upon which this report is based was done under a cooperative agreement between the U.S. Energy Research and Development Administration, Laramie Energy Research Center, and the University of Wyoming. Reference to specific trade names or manufacturers does not imply endorsement by the U.S.E.R.D.A. Prior to completion of this work Laramie Energy Research Center was a unit of the U.S. Bureau of Mines.

# Sulfur Compounds in Oils From the Western Canada Tar Belt

D. M. CLUGSTON, A. E. GEORGE, D. S. MONTGOMERY, G. T. SMILEY, and H. SAWATZKY

Fuels Research Centre, c/o 555 Booth Street, Ottawa, Canada K1A OG1

*Sulfur compounds in the gas oil fractions from two bitumens (Athabasca oil sand and Cold Lake deposit), a heavy oil (Lloydminster) from Cretaceous reservoirs along the western Canada sedimentary basin, and a Cretaceous oil from a deep reservoir that may be mature (Medicine River) are investigated. The gas oil distillates were separated to concentrates of different hydrocarbon types on a liquid adsorption chromatographic column. The aromatic hydrocarbon types with their associated sulfur compounds were resolved by gas chromatographic simulated distillation and then by gas solid chromatography. Some sulfur compounds were further characterized by mass spectrometry. The predominant sulfur compounds in these fractions are alkyl-substituted benzo- and dibenzothiophenes with short side chains which have few dominant isomers.*

In this work the sulfur compounds in the gas oil of three Cretaceous heavy oils from the edge of the Alberta sedimentary basin were investigated. These crude oils were obtained from the Athabasca, Cold Lake, and Lloydminster deposits and are believed (1) to belong to the same oil system which implies like modes of origin. The geographic location of these deposits is shown in Figure 1. This investigation was conducted to develop the analytical capability used to follow the maturation of the sulfur compounds in these oils (2).

### Experimental

**Samples.** The following samples were investigated:

1. The pentane extract from the Athabasca bituminous sand obtained from the quarry of Great Canadian Oils Sands, Ltd.

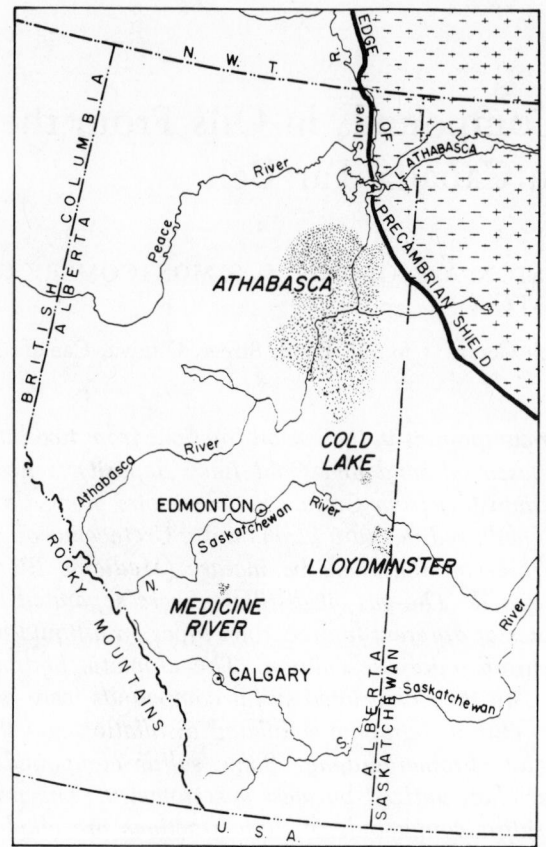

*Figure 1.   The Western Canada sedimentary basin*

2. Cold Lake bitumen obtained by steam injection at 1500 ft (457 m) from the Mannville Pool by Imperial Oil Co.

3. Lloydminster oil produced by Husky Oil Co. from the Sparky Formation location 9A-3-50-1 (W4), depth 1893 ft (577 m).

4. Medicine River oil produced by Hudson Bay Oil and Gas Co. from the Glauconite zone, location 2-17-39-4W5M, depth 7326 ft (2233 m).

**Molecular Distillation.** The pentane extract from the Athabasca sands and the other oils without any treatment were distilled in an Arthur F. Smith molecular still at pressure ranges of 75–250$\mu$ and temperatures of 225–230°C. The light ends were obtained in the cold trap between the still and the vacuum pump.

**Chromatographic Separations.** The distillates from the oils were separated first by liquid chromatography, then by gas chromatographic-simulated distillation according to boiling point, and finally by gas chromatography on lithium chloride-coated silica, as shown in Figure 2.

**Liquid Chromatography.** The distillates were separated on a dual-packed (silica gel–alumina gel) chromatographic column according to the procedure developed by the U.S. Bureau of Mines in conjunction with API project 60 (3). The oil fractions were obtained by collecting 200 ml fractions of the eluent and evaporating the solvents by warming and using streams of nitrogen. Various fractions of saturated, mono-nuclear aromatic, binuclear aromatic, polynuclear aromatic hydrocarbons, and polar materials were obtained.

**Gas Chromatographic-Simulated Distillation.** The fractions obtained from liquid chromatography were gas chromatographed on nonpolar silicone rubber, SE–30 (5 ft U-shaped glass column of 4 mm I.D.) to achieve separations according to boiling points. Normal alkanes were used as reference compounds. The column effluent was split between a flame ionization detector and a trapping out assembly in a ratio of 1:9. Each collecting tube contained 10 mg of Chromosorb W. The helium carrier gas had a flow rate of 175 ml/min. Sulfur chromatograms were obtained with a Dohrmann microcoulometric quantitative detector which was also used to determine total sulfur contents of the samples. The carrier gas flow rate was 75 ml/min when the coulometric detector was used.

**Gas–Solid Chromatography and Mass Spectrometry.** The cuts trapped out from the simulated distillations were rechromatographed on lithium chloride-coated silica columns (4). The column effluent was split with a portion directed to a flame ionization detector and the other to the mass spectrometer in a ratio of 1:4. This chromatographic step greatly facilitated the interpretation of the mass spectra. In many cases it appeared as though pure compounds were obtained. Only some of these gas–solid chromatograms will be discussed.

The mass spectrometer was a CEC 21–104 equipped with a Watson–Biemann type G.C. interface. The source was maintained at 250°C. An

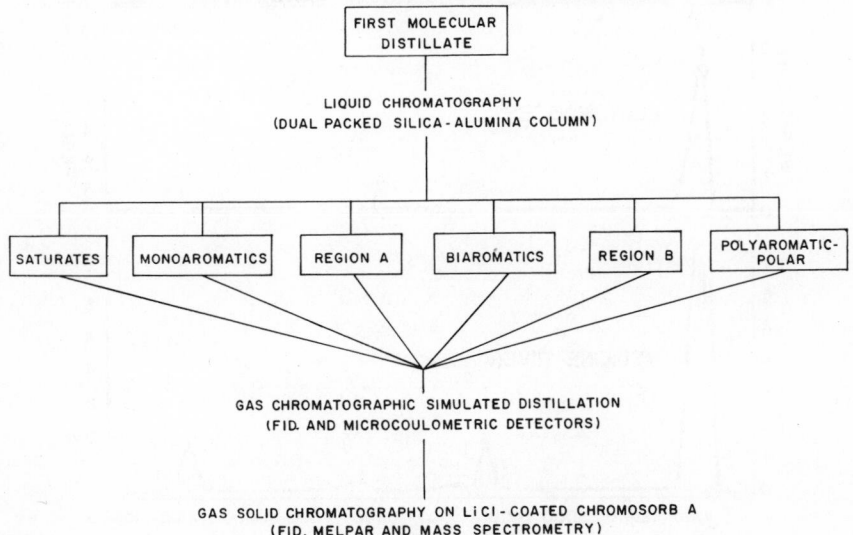

*Figure 2.    Chromatographic separations*

### Table I.  Distillate Yields and Their Sulfur Contents in the Oils of Alberta Tar Belt

|  | Athabasca | Cold Lake | Lloyd-minster | Medicine River |
|---|---|---|---|---|
| Light ends (wt %) | 4.5 | 5.9 | 12.5 | 44.8 |
| First distillate (wt %) | 37.5 | 31.0 | 36.0 | 43.5 |
| Sulfur content (wt %) | 2.90 | 2.85 | 2.50 | 0.80 |

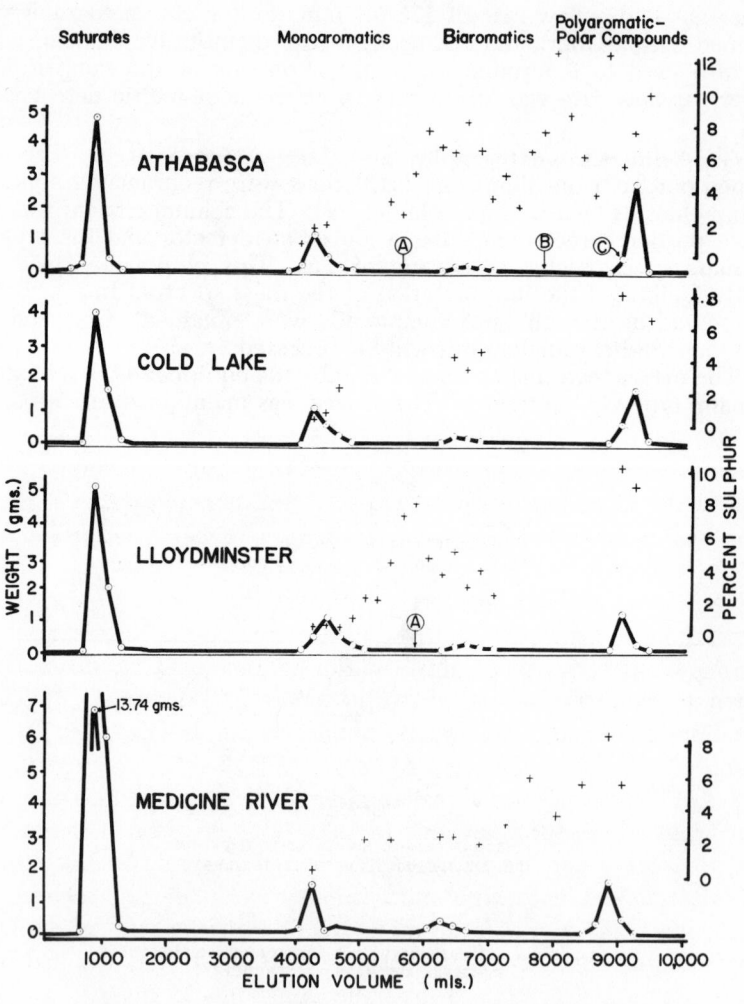

Figure 3.  Liquid–solid chromatographic separations

ionization potential of 70 eV was used throughout with a trap current of approximately 80 $\mu$ amp. The spectra were scanned electrically.

## Results and Discussion

Table I shows the molecular distillation yields and distribution of the total sulfur contents of the first distillate in the four oils.

The liquid chromatograms of the four gas oils are shown in Figure 3. The sulfur content of some of the fractions is also shown (denoted by + symbols). The proportions of the hydrocarbon types of the oils are shown in Table II.

### Table II.   Hydrocarbon Type Distribution

|  | Athabasca | Cold Lake | Lloyd-minster | Medicine River |
|---|---|---|---|---|
| Saturates (wt %) | 37.1 | 42.6 | 50.7 | 69.6 |
| Monoaromatics (wt %) | 21.3 | 22.9 | 22.7 | 13.0 |
| Biaromatics (wt %) | 15.8 | 14.1 | 13.1 | 7.1 |
| Polyaromatics (wt %) | 25.6 | 20.5 | 13.6 | 10.3 |

There is an increasing trend in the amount of saturates in the sequence Athabasca, Cold Lake, Lloydminster, and Medicine River. This trend is reversed for the biaromatic and especially for the polyaromatic–polar fraction.

The saturated hydrocarbon fractions contain traces of sulfur. The gas–solid chromatography of the Lloydminster saturates boiling in the range of the $C_{25}$ normal alkane, using a Melpar flame photometric detector, shows that the sulfur compounds are retained much longer than the hydrocarbons. This is typical for alkyl sulfides.

There was a small amount of sulfur in the mononuclear aromatic hydrocarbon fractions, which did not show any appreciable resolution on the microcoulometric trace obtained during simulated distillation. According to the U.S. Bureau of Mines report (3) describing this type separation, thiophenes and cyclic sufides could be expected in this fraction.

Sulfur compounds in the main binuclear aromatic fractions as well as in the main polynuclear aromatic hydrocarbon and polar material fractions from the liquid chromatography were investigated. Also studied were the fractions with high sulfur content in the region between the largest mono- and binuclear aromatic fractions, designated by the letter A on the liquid chromatograms; between the largest binuclear and polynuclear aromatic fractions, B, and the polyaromatic fraction, C (Figure 3).

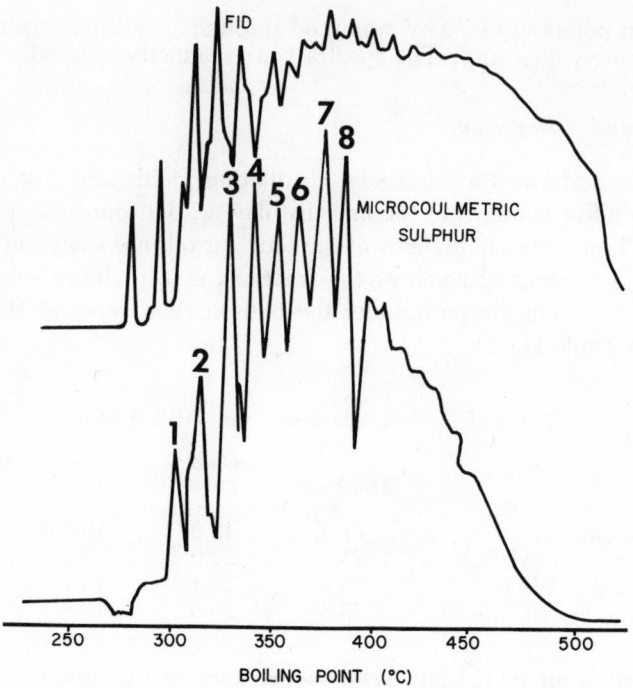

*Figure 4.   Lloydminster fraction A, SE-30 on Chromosorb*
*W*

In Figures 4 and 5 the microcoulometric sulfur peaks of the fractions in region A are well resolved and appear at regular intervals as would be expected for a homologous series of compounds. Most of the sulfur peaks have matching flame ionization peaks. Because the chromatograms of the fractions from region A of both Athabasca and Lloydminster oils are similar, it was decided to study only one of them further by mass spectroscopy. The Lloydminster oil was selected since it has more material in the lower molecular weight range.

The material represented by the sulfur peaks 1–4 were trapped out individually, using the appropriate flame ionization peaks and shoulders that matched the sulfur peaks and then rechromatographed on the lithium chloride-coated silica. The trapped-out cuts represent material having boiling ranges of 275°–296°C, 296°–309°C, and 309°–320°C respectively. The resolved material was studied by mass spectroscopy. Only the data thought to be related to the sulfur compounds are discussed. The flame ionization and Melpar flame photometric sulfur chromatograms of the first three of these trapped-out cuts are shown in Figures 6, 7, and 8. Although the Melpar sulfur trace is not quantitative, whenever matching predominant peaks were obtained on both the flame ionization and the sulfur traces, they were considered to represent substantial sulfur-containing material.

The mass spectrum obtained from the material eluting at 153°C in Figure 6 contained a molecular ion of $m/e$ 190 and intense fragment ions with $m/e$ 175 and 147, corresponding to a four-carbon substituted benzothiophene. Spectra taken at higher temperatures in the chromatogram also contained these same ions but in different relative intensities. High relative intensity of the molecular ion and M–15 ion is taken to mean that the side chains on the aromatic nucleus are short—probably methyl —while low relative intensity of M and M–15 relative to the ion corresponding to the aromatic nucleus plus one methylene group (*i.e.*, a tropyllium type of structure) indicates fewer and longer side chains.

The material eluting at 162°C (Figure 7) produced ions at $m/e$ 204, 189, and 147 corresponding to a five-carbon substituted benzothiophene. As before, the early eluting material appeared to have one long side chain while later in the chromatogram, material having more and somewhat shorter side chains appeared. This trend appeared in all chro-

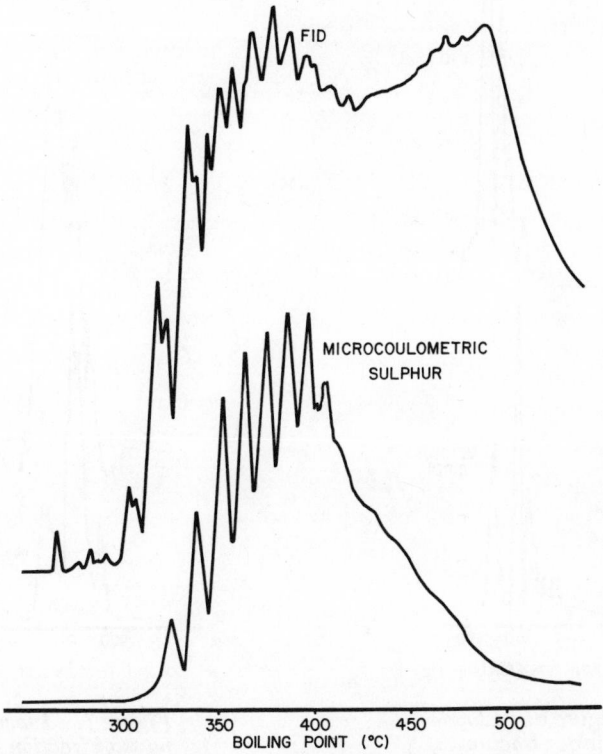

*Figure 5. Athabasca fraction A, SE-30 on Chromosorb W*

matograms. The presence of molecular ions of $m/e$ 202 and 206 in some
of these spectra may indicate the presence of cycloalkyl benzothiophenes
and thiaindanes respectively.

In subsequent cuts from Lloydminster fraction A it appeared that
benzothiophenes with six alkyl carbon atoms for cut 3 and seven for cut 4
were involved. On the basis of this trend, it is assumed that each successive sulfur peak, obtained during simulated distillation, as shown in
Figures 4 and 5, represents benzothiophene with an additional alkyl
carbon atom. Thus dominant benzothiophene peaks are obtained with as
many as 11 substituting carbon atoms in peak 8. Then the amounts of
sulfur compounds involved in these chromatograms decrease markedly
which is not the case for the accompanying hydrocarbons, as shown by
the flame ionization trace. High resolution mass spectral data appear to

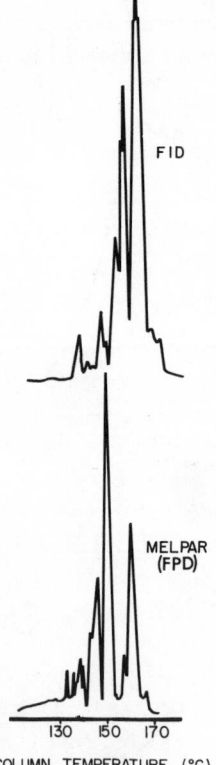

Figure 6.  Lloyd-
minster fraction A,
LiCl on Chromo-
sorb A, boiling
range 275°–296°C

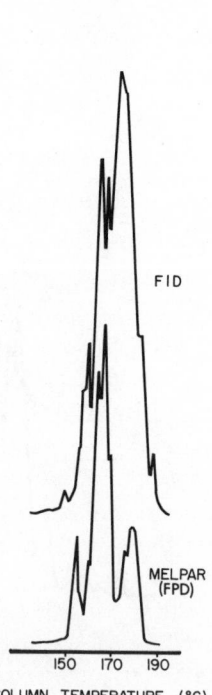

Figure 7.  Lloyd-
minster fraction A,
LiCl on Chromo-
sorb A, boiling
range 296°–309°C

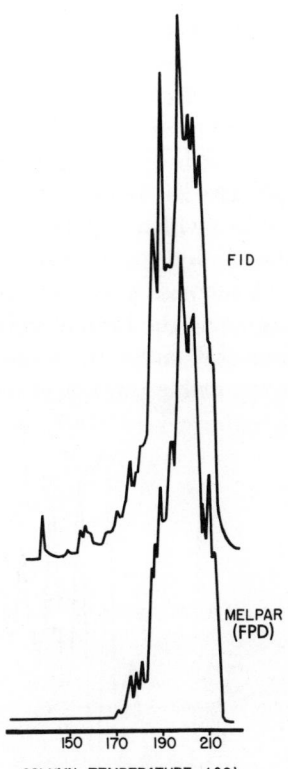

FID

MELPAR
(FPD)

*Figure 8.   Lloydmin-*
*ster fraction A, LiCl on*
*Chromosorb A, boiling*
*range 309°–320°C*

150   170   190   210
COLUMN TEMPERATURE ( °C )

support the assumption that these compounds are benzothiophenes (5).

The microcoulometric sulfur chromatograms that were obtained during simulated distillation of the largest biaromatic fractions of the four
oils are shown in Figures 9–12. The dominant peaks are well resolved
and appear at regular intervals similar to those from a homologous series
of compounds. The fractions of all the four oils were studied by mass
spectrometry. In the case of the Athabasca fraction, which had a high
sulfur content, the sulfur peaks had matching flame ionization peaks. It
seems that the same sulfur compounds appear in all corresponding fractions of the four oils.

The sulfur compounds appear to be alkyl benzothiophenes. The
mass spectra of these benzothiophenes showed prominent molecular ions
and quite limited fragmentation as though most of the alkyl groups are
methyls. This is in contrast to the benzothiophenes with longer alkyl
groups found in the small fractions with high sulfur content that were
obtained in region A. It would be expected that the benzothiophenes
with shorter but more alkyl groups would be retained longer on the elec-
trophylic adsorbents than those with longer but fewer electron donating

groups. The fact that most of the alkyl substituents on the benzothiophenes in the largest biaromatic fraction are methyls accounts for the
limited number of dominant isomers that appear in these fractions as is
shown by the well resolved peaks.

The chromatogram in Figure 9 of the largest biaromatic fraction
from the Athabasca oil shows three dominant well resolved peaks, num-
bered 1, 2, and 3, which appear to represent benzothiophenes with five,
six, and seven methylene groups with molecular ions of $m/e$ 204, 218,
and 232. With increasing molecular weight, the amounts of sulfur compounds decrease and the resolution becomes poorer as would be expected
if one assumes that larger substituting alkyl groups are present causing
the number of possible isomers to increase.

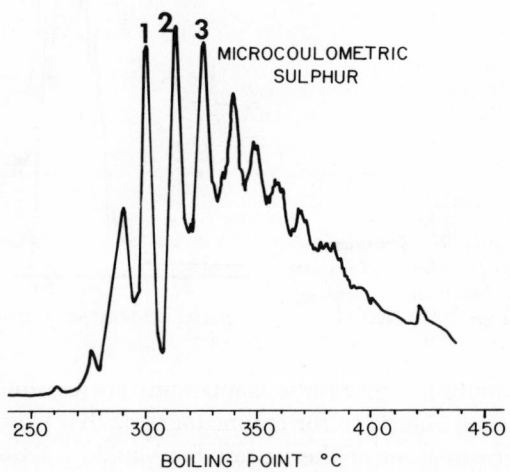

*Figure 9.    Athabasca biaromatics fraction, SE-
30 on Chromosorb W*

The chromatogram from the corresponding Lloydminster fraction,
Figure 10, is similar to that from the Athabasca but appears to contain
considerably more lower molecular weight benzothiophenes, particularly
those with four methylene groups represented by peak 2. The Cold Lake
fraction, Figure 11, is not well resolved but also shows peaks that match
those from the other oils.

The chromatogram of the corresponding fraction from the Medicine
River oil, Figure 12, shows the best peak resolution, and these peaks
appear to represent the same benzothiophenes present in the other oils.
This is believed to be a thermally mature oil, and it is therefore expected
that the alkyl side chains would be reduced to methyl residues. The

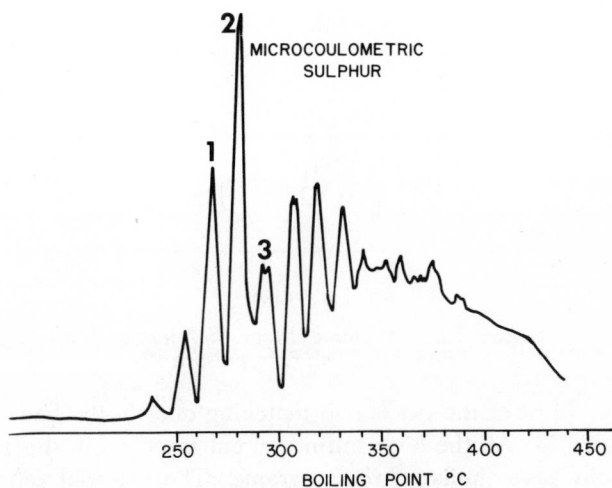

*Figure 10. Lloydminster biaromtics fraction, SE-30 on Chromosorb W*

alkyl naphthalenes that accompany the benzothiophenes also appear to possess mostly methyl substitution.

Before leaving the sulfur compounds in the biaromatic fraction, it should be mentioned that only the material with well-resolved sulfur peaks was examined by mass spectra. Other sulfur compounds besides benzothiophenes possibly might be present in the poorly resolved higher molecular weight materials.

The small fractions with high sulfur content that were obtained in the region B of the liquid chromatograms also gave sulfur gas chromatograms with fairly well-resolved peaks in the sulfur traces of the simulated

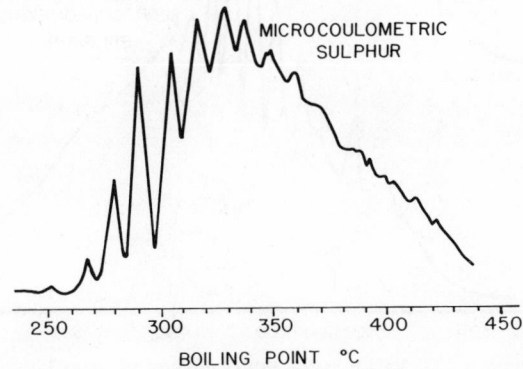

*Figure 11. Cold Lake biaromatics fraction, SE-30 on Chromosorb W*

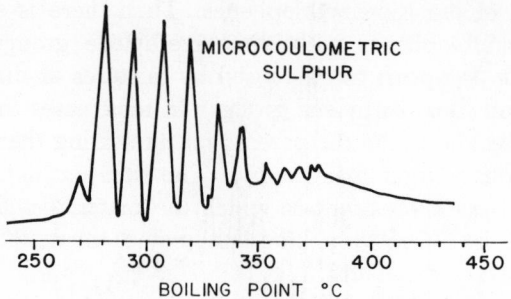

*Figure 12. Medicine River biaromatics fraction, SE-30 on Chromosorb W*

distillation. Most of the peaks had matching ones in the flame ionization trace, Figure 13. All the high sulfur content fractions in this region B of the four oils gave similar chromatograms. The material represented at the front end of the chromatogram is not very well resolved. The first recognizable peak, 1, appears to be caused by dibenzothiophenes with mainly two methylene groups. The $m/e$ ratio was 212, and the only fragment ion was at $m/e$ 197 which is the lowest mass fragment found in this series. It is assumed to represent the aromatic nucleus plus one alkyl

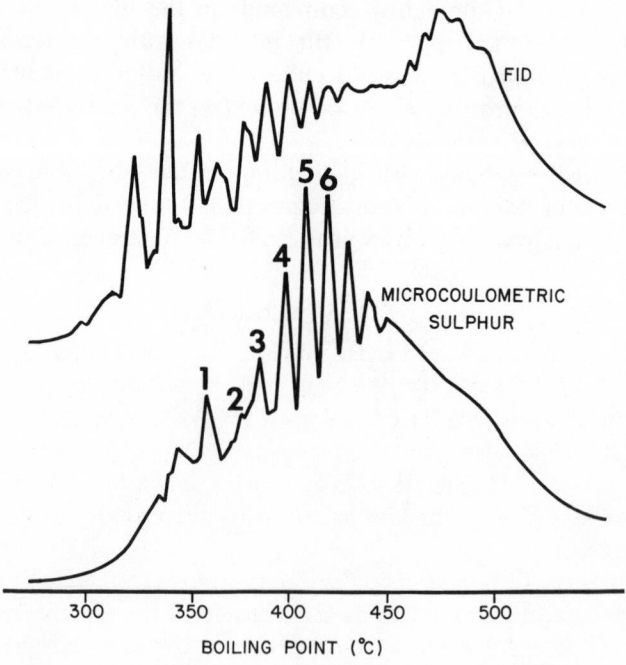

*Figure 13. Athabasca fraction B, SE-30 on Chromosorb W*

carbon atom of the dibenzothiophenes. Then there is a small peak, 2, where dibenzothiophenes with three methylene groups appear to be present. Peak 3 appears to be caused by mixtures of dibenzothiophenes with three and four methylene groups. In most cases the fragment ions were more abundant than the parent ions, indicating that the compounds had alkyl groups larger than methyl. The situation might be similar to that for the benzothiophenes in which the material with longer chains was eluted in the liquid chromatography before the material that contains more but shorter side chains.

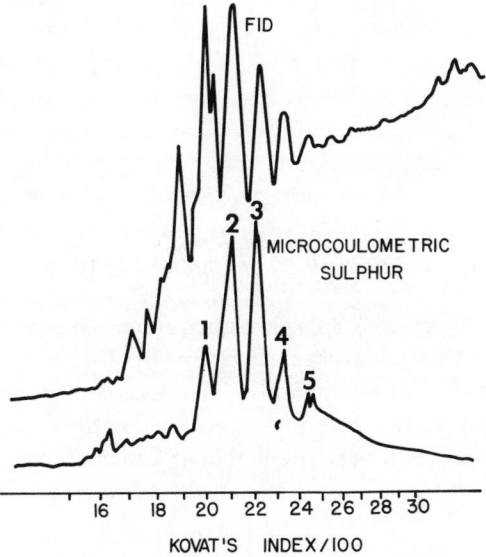

*Figure 14. Athabasca fraction C (polyaromatics) on Chromosorb W*

In this fraction B, mass spectra were not obtained for material of molecular weight higher than the dibenzothiophenes with four methylene groups. It is inferred that the two most prominent peaks, 5 and 6, can be ascribed to dibenzothiophene with six and seven methylene groups. With increasing molecular weight the peaks become less prominent, and the amount of sulfur compounds decreases. As can be seen in Figure 13, this is quite different from the hydrocarbons that accompany these sulfur compounds.

The first polyaromatic polar fraction from the Athabasca oil was similar to the largest fraction in this class from the Lloydminster oil. All the simulated distillation chromatograms of the first polyaromatic polar fractions in the C region of the liquid chromatograms are shown in Figures 14–17. Sulfur peaks from the three oils are fairly well resolved,

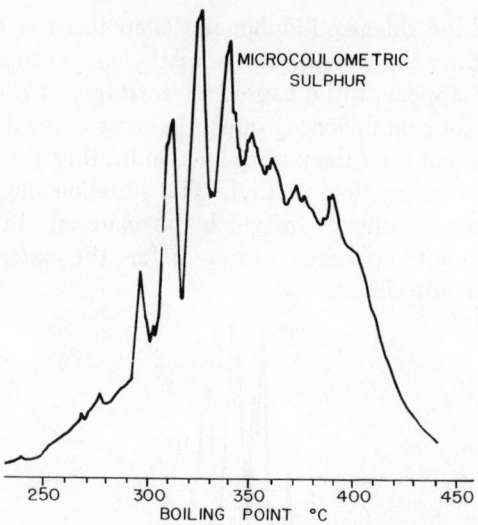

*Figure 15. Lloydminster polyaromatics fraction, SE-30 on Chromosorb W*

at least in the lower molecular weight region. It seems that the initial portion of the Athabasca chromatogram, up to Kovat's Index of 2400, is due mostly to sulfur compounds. The main sulfur compounds appear to be dibenzothiophenes with three and four methylene groups which are represented by the two very predominant peaks, 2 and 3 with molecular

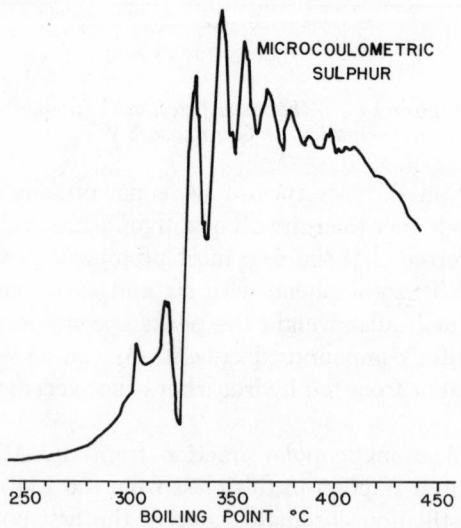

*Figure 16. Cold Lake fraction C (polyaromatics), SE-30 on Chromosorb W*

ions of $m/e$ 226 and 240. These dibenzothiophenes constitute the major portion of the sulfur compounds in this fraction. Some dibenzothiophenes with two, five, and six methylene groups represented by peaks 1, 4, and 5 respectively, also appear in much smaller quantities.

The predominant sulfur peaks in the Lloydminster and Cold Lake fractions appear from their mass spectra to be the same as in the Athabasca although they contain more of the unresolved higher molecular weight material. In the Medicine River fraction we could not find sufficiently intense ions that could be ascribed to dibenzothiophenes as in the other oils. The chromatogram, Figure 17, shows the sulfur response increasing progressively with boiling point which indicates that the major portion of the sulfur compounds are of substantially higher molecular weight than those in the other oils.

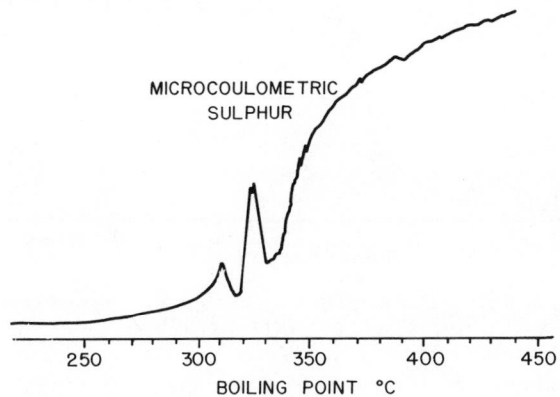

MICROCOULOMETRIC
SULPHUR

250    300    350    400    450
BOILING   POINT   °C

*Figure 17.   Medicine River fraction C (polyaromatics), SE-30 on Chromosorb W*

The chromatogram of the largest polyaromatic and polar fraction from the Athabasca gas oil can be seen in Figure 18. This fraction has a sulfur content of more than 7.7% within the distillation range up to a boiling point of 500°C. Assuming the presence of one sulfur atom per molecule, about half the fraction consists of sulfur compounds. Some of the ill-defined sulfur peaks appear, together with matching flame ionization peaks, at intervals in the boiling point suggesting a difference of a methylene group.

According to the mass spectra, alkyl substituted dibenzothiophenes or possibly naphthothiophenes seem to be quite prominent in this fraction, with the substitution ranging from two to eight methylene groups. It seems likely that few of the alkyl groups would be larger than methyl, so the poor resolution of sulfur compounds probably is caused by other types of sulfur compounds in addition to dibenzothiophenes, such as

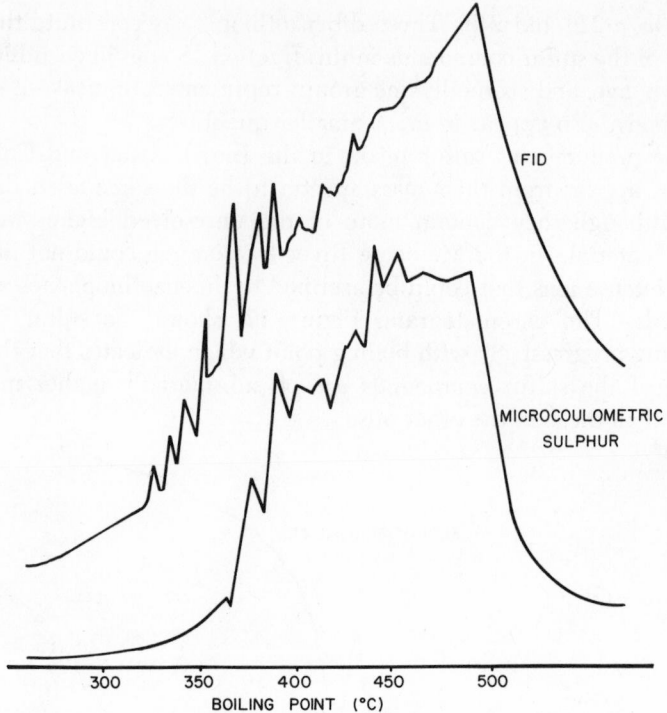

*Figure 18. Athabasca main polyaromatic–polar fraction,*
*SE-30 on Chromosorb W*

naphthothiophenes. In the mass spectra there were abundant ions that might rise from naphthobenzothiophenes or phenylbenzothiophenes. The presence of small amounts of alkyl diphenylsulfides is also possible.

### Conclusion

The dominant sulfur compounds in the gas oils of the heavy oil and bitumens are alkyl substituted benzo- and dibenzothiophenes, with a limited number of isomers. While some of these thiophenes have long side chains, those in the major fractions have short side chains. The dominance of short side chains might be ascribed to the maturation effects. The benzothiophenes in the deeper Medicine River oil are similar to those from the heavy oils, however, the same dibenzothiophenes are not found in significant amounts.

### Acknowledgment

We are grateful to the Husky Oil Co. Ltd. for a sample of Lloyd-minster oil and to the Hudson Bay Oil and Gas Co. Ltd. for a sample of

Medicine River oil. We wish to thank T. M. Potter and R. M. Evis for technical assistance.

*Literature Cited*

1. Vigrass, Laurence W., *Am. Assoc. Pet. Geol., Bull.* (1968) **52**, 1984.
2. Montgomery, D. S., Clugston, D. M., George, A. E., Smiley, G. T., Sawatzky, H., "Oil Sands Fuel of the Future," p. 168, Canadian Society of Petroleum Geologists, Calgary, Alta., Sept., 1974.
3. Hirsh, D. E., Hopkins, R. L., Coleman, H. J., Cotton, F. O., Thompson, C. J., *Anal. Chem.* (1972) **44**, (6) 915.
4. Sawatzky, H., George, A. E., Smiley, G. T., *Amer. Chem. Soc. Petroleum Preprints* (1973) **18**, 99.
5. Sawatzky, H., Smiley, G. T., George, A. E., Clugston, D. M., *Amer. Chem. Soc. Fuels Preprints* (1971) **15**, 83.

RECEIVED December 16, 1974.

# 3

# Characterization of Synthetic Liquid Fuels

R. G. RUBERTO, D. M. JEWELL,[1] R. K. JENSEN, and D. C. CRONAUER

Gulf Research and Development Co., Pittsburgh, Pa. 15230

*Analytical separation and spectroscopic techniques normally used for petroleum crudes and residues were modified and used to characterize coal liquids, tar sands bitumens, and shale oils. These techniques include solvent extraction, adsorption, ion-exchange, and metal complexing chromatography to provide discrete fractions. The fractions are characterized by various physical and spectroscopic methods such as GLC, MS, NMR, etc. The methods are relatively fast, require only a few grams of sample, provide compound type fractions for detailed characterization, and provide comparative compositional profiles for natural and synthetic fuels. Additional analytical methods are needed in some areas.*

In order to decide what is the best use of a fuel—natural or synthetic—and/or which one out of a number of fuels is the best for a specific application, it is necessary to know the characteristics of each fuel and to be able to compare the characteristics of one fuel directly with those of another.

Experience has shown that to obtain meaningful results in analyzing petroleum crudes and residues, it is necessary to separate a sample into a certain number of well-defined fractions and to analyze these fractions in detail. Conclusions as to the composition of the original sample are then made by combining the results of the analyses on each fraction in a manner consistent with the steps performed to obtain them. This same approach is used for the synthetic liquid fuels, and this paper reports our results obtained on such materials.

The literature on tar sands and tar sand bitumen is not very rich as compared with that on shale oils and coal. Most of the currently available

[1] Present Address: Ford Motor Co., P.O. Box 2053, Dearborn, MI. 48121

data are limited to the material of the Athabasca deposit in the Province of Alberta, Canada, primarily because of its size and location. Detailed general information on the Athabasca tar sands and tar sand bitumen is available in Refs. *1* and *2*. In addition, the analysis of this bitumen has been reported by various workers (*3, 4, 5, 6, 7*).

The literature on oil shale and shale oils is much richer and goes back several decades. However, because of its size and location, most of the literature deals only with one oil shale formation—the Green River Formation.

Space limitations prohibit a literature survey here; however, much information on oil shales and shale oils in general is available (*8, 9, 10, 11*). The remainder of the available data can be divided into two general categories: data produced by various retorting methods with the purpose of obtaining a fuel (*12–23*) and data produced by solvent extraction methods for geological and geochemical studies (*24–35*).

An enormous amount of data dealing with all aspects of the chemical and physical properties of coal is available. However, only four reviews have been referenced which are particularly comprehensive (*36, 37, 38, 39*).

## Experimental

**Samples Preparation.** The coal liquids were derived from the catalytic liquefaction of Pittsburgh Seam bituminous and Wyoming sub-bituminous coals. The analysis of these coals is given in Table I. The coals were liquefied in a bench-scale catalytic unit using cyclone overhead product as recycle solvent to insure that the liquid products were derived from the coal and not the solvent. The product streams from

### Table I. Analysis of Coal Samples

| Coal Sources<br>Rank | Pittsburgh Seam<br>Bituminous | Big Horn, Wyo.<br>Sub-bituminous |
|---|---|---|
| *Proximate analysis (wt. %)* | | |
| Moisture | 2.5 | 19.6 |
| Volatile matter | 33.4 | 34.0 |
| Fixed carbon | 57.4 | 41.2 |
| Ash | 6.7 | 5.2 |
| *Chemical analysis (wt. %) (moisture free basis)* | | |
| Carbon | 78.68 | 69.15 |
| Hydrogen | 4.96 | 4.69 |
| Nitrogen | 1.57 | 1.23 |
| Oxygen (difference) | 6.29 | 17.75 |
| Sulfur | 1.65 | 0.72 |
| Ash | 6.87 | 6.46 |

the unit consisted of gases, water, light ends typically boiling in the range of 150–500°F, and slurry. The slurry product was filtered to remove the undissolved coal plus mineral matter prior to analysis. Based on a calculated material balance for processing 1.0 ton of Pittsburgh Seam coal, the light ends and filtrate yields were 135 and 1350 lbs, respectively. Similarly, yields of light ends and filtrate from the sub-bituminous coal were 265 and 740 lbs, respectively. The term "coal liquids" refers to the filtered slurry product with a boiling point above 130°F.

A sample of raw bitumen recovered from Athabasca tar sands and provided by Sun Oil Co. was analyzed without further upgrading. Three distillate cuts of shale oil obtained from The Oil Shale Corp. were also analyzed without further upgrading.

**Separation Into Fractions.** The separation procedure, developed for petroleum crudes and residues (*40*) is illustrated in Figure 1. This procedure is applicable to samples having a boiling point higher than 470°F.

The distillation step is necessary only with samples containing low boiling materials. These materials must be removed since they would be lost during the subsequent steps in which solvents are used and then evaporated to recover the fractions.

The material boiling above 470°F is separated into oils, resins, and *n*-pentane insoluble residue. The residue is separated into asphaltenes and benzene insolubles by extraction with benzene while the oils are separated into aromatics and saturates. The saturates can be further separated into *n*-paraffins and non-*n*-paraffins with 5 Å molecular sieves

| | | Table II. Separation | |
| | | Coal Liquids | |
| Sample Description | Tar Sands Bitumen | Big Horn Coal | Pitt Seam Coal |
|---|---|---|---|
| Boiling range (°F) | 600+ | 130+ | 130+ |
| Fraction of total sample | 100.0 | 100.0 | 100.0 |
| Distillate (BP <470°F) | 0.0 | 28.3 | 25.2 |
| saturates[a] | — | 14.0 | 14.2 |
| aromatics[a] | — | 14.2 | 10.8 |
| olefins[a] | — | 0.1 | 0.2 |
| Residue (BP >470°F) | 100.0 | 71.7 | 74.8 |
| saturates | 16.5 | 3.6 | 1.8 |
| aromatics | 47.8 | 58.3 | 36.2 |
| monoaromatics | 7.0 | 9.0 | 5.9 |
| di- + triaromatics | 30.1 | 42.0 | 26.7 |
| polyaromatics | 10.7 | 7.3 | 3.6 |
| resins (non-hydrocarbons) | 25.9 | 8.3 | 11.1 |
| asphaltenes | 9.8 | 1.5 | 20.3 |
| benzene insolubles | <0.1 | <0.1 | 5.5 |

[a] By FIA, ASTM D 1319.

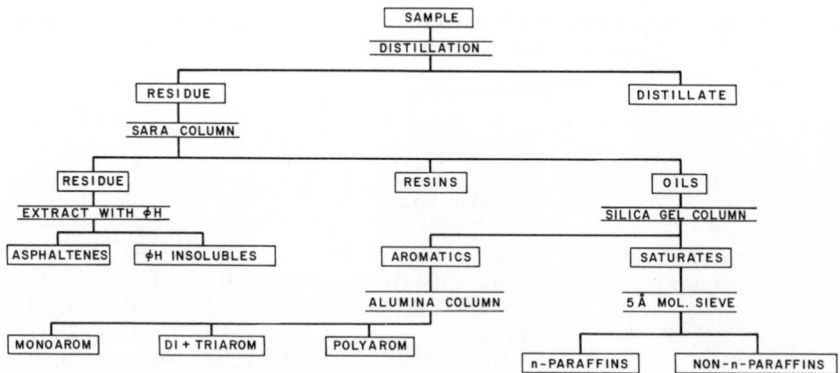

*Figure 1.   Separation scheme*

(41) while the aromatics are separated into three additional fractions on alumina (42). The results of the separations are listed in Table II.

**Analysis Methods.** Unless otherwise specified, the average molecular weights were obtained by vapor pressure osmometry (VPO) in benzene according to ASTM D2503. For the aromatic fractions, an average molecular weight is also obtained by $^1$H NMR and reported in the respective tables.

Carbon and hydrogen were determined by a microcombustion method and nitrogen by a micro-Kjeldahl method. The oxygen in the total samples was determined by neutron activation and that in the fractions by

**Data (Wt %)**

| | Shale Oils | |
|---|---|---|
| *Cut I* | *Cut II* | *Cut III* |
| 140–540 | 540–680 | 660+ |
| 32.6 | 10.4 | 57.0 |
| 66.5 | 2.7 | 0.0 |
| 14.3 | — | — |
| 33.9 | — | — |
| 18.3 | — | — |
| 33.5 | 97.3 | 100.0 |
| 6.0 | 31.0 | 12.3 |
| 14.7 | 27.0 | 15.8 |
| 8.3 | 8.0 | 5.3 |
| 4.2 | 13.4 | 9.3 |
| 2.3 | 5.6 | 1.3 |
| 12.1 | 38.6 | 68.6 |
| 0.7 | 0.8 | 3.3 |
| <0.1 | <0.1 | <0.1 |

**Table III. Chemical Analysis and Molecular Weights of Tar Sands Bitumen, Coal Liquids, and Their Fractions**

| | Molecular Weight | C (wt %) | H (wt %) | N (wt %) | O (wt %) | S (wt %) |
|---|---|---|---|---|---|---|
| Tar sands bitumen | — | 82.98 | 10.42 | 0.42 | 1.15 | 4.60 |
| saturates | 365 | 86.00 | 14.00 | — | — | — |
| aromatics | 460 | — | — | — | — | — |
| monoaromatics | 360 | 88.55 | 11.36 | — | — | — |
| di- + triaromatics | 365 | 85.04 | 9.45 | 0.02 | 1.14 | 3.80 |
| polyaromatics | 1,400 | 79.36 | 9.57 | 0.42 | 3.40 | 6.89 |
| resins | 1,300 | 81.15 | 9.04 | 1.34 | 3.35 | 5.31 |
| asphaltenes | 5,100 | 78.84 | 7.80 | 1.19 | 4.53 | 8.46 |
| Big Horn coal liquids | — | 89.18 | 8.97 | 0.40 | 1.03 | 0.04 |
| saturates | 300 | 86.12 | 13.65 | — | — | — |
| aromatics | 222 | — | — | — | — | — |
| monoaromatics | 285 | 88.09 | 10.10 | 0.06 | 1.82 | 0.00 |
| di- + triaromatics | 220 | 92.38 | 7.13 | 0.01 | 0.80 | 0.15 |
| polyaromatics | — | 84.19 | 6.60 | 0.20 | 7.80 | 0.97 |
| resins | 380 | 83.84 | 7.09 | 1.62 | 7.15 | 0.30 |
| asphaltenes | — | 87.37 | 6.06 | 1.25 | 4.92 | 0.62 |
| benzene insolubles | — | — | — | — | — | — |
| Pitt seam coal liquids | — | 89.05 | 8.18 | 0.82 | 1.47 | 0.17 |
| saturates | — | 85.40 | 14.17 | — | — | — |
| aromatics | 240 | — | — | — | — | — |
| monoaromatics | 290 | — | — | — | — | — |
| di- + triaromatics | 235 | 92.52 | 7.20 | 0.01 | 0.67 | 0.35 |
| polyaromatics | — | — | — | 0.05 | 7.75 | 0.34 |
| resins | 440 | 81.30 | 7.33 | 1.37 | 5.77 | 0.42 |
| asphaltenes | 775 | 87.73 | 6.86 | 1.76 | 3.92 | 0.38 |
| benzene insolubles | — | 85.87 | 5.46 | 2.12 | 5.64 | 0.57 |

a modified Unterzaucher method (43). Sulfur was determined by a combustion method similar to ASTM D1552. Tables III and IV report all the molecular weights and elemental data.

The simulated distillation data (Table V and Figures 2, 4, 6) and the FIA analyses of the distillates (Table II) were obtained by standard ASTM methods D2887 and D1319, respectively. The mass spectrometric analyses (MS) of the saturates fractions (Table VI) were obtained by an in-house method similar to that of Hood and O'Neal (44). The aromatic fractions were analyzed by the proton nuclear magnetic resonance (NMR) method of Clutter et al. (45), and the results are reported in Tables VII and VIII.

*Discussion of Results*

**Tar Sand Bitumen.** This material is very viscous, black, and contains a considerable amount of sulfur, some nitrogen and oxygen, and no light ends (*see* Tables II and III). The saturate fraction amounts to about 17% and is a clear colorless liquid with an average molecular weight of 365 (Table III). The carbon number ranges from about 14 to well above 44 (Figure 2), and its final boiling point is above 1000°F (Table V). These characteristics are reconciled by the observation that no alkanes are present as shown by the mass spectrometric data (Table VI), the fact that no material was removed when this fraction was treated with 5 Å molecular sieves, and the lack of sharp peaks on the gas–liquid chromatogram (GLC) of this fraction (Figure 2). The inability to remove alkanes from this type of material was also reported by Speight (*4*). The condensed and noncondensed cycloalkanes are evenly distributed, and more than half of the condensed cycloalkanes have only two rings, as indicated by mass spectrometry (Table VI).

Table IV. Chemical Analysis and Molecular Weights of Shale Oil Fractions

|  | Molecular Weight | C (wt %) | H (wt %) | N (wt %) | O (wt %) | S (wt %) |
|---|---|---|---|---|---|---|
| Cut I, total | 145[a] | 85.30 | 12.12 | 0.98 | 0.98 | 0.80 |
| distillate | 130[a] | 85.99 | 12.50 | 0.63 | 0.99 | 0.75 |
| saturates | 230[a] | 85.84 | 14.31 | — | — | — |
| aromatics | 170[b] | 85.44 | 10.09 | 0.25 | 2.17 | 2.05 |
| monoaromatics | 240[b] | 83.26 | 11.65 | <0.01 | 4.80 | <0.01 |
| di- + triaromatics | 190[b] | 86.35 | 8.27 | 0.02 | 5.06 | <0.01 |
| polyaromatics | 200[b] | 76.89 | 9.76 | 0.35 | 10.52 | 2.46 |
| resins | 445 | 75.89 | 8.38 | 1.28 | 10.82 | 3.44 |
| Cut II, total | 260 | 84.68 | 11.07 | 2.10 | 1.23 | 0.68 |
| saturates | 266[a] | 85.99 | 14.34 | — | — | — |
| aromatics | 325 | — | — | — | — | — |
| monoaromatics | 390 | 86.61 | 12.07 | <0.01 | 1.67 | <0.01 |
| di- + triaromatics | 290 | 87.79 | 9.26 | <0.01 | 1.70 | 1.84 |
| polyaromatics | — | 78.97 | 9.58 | 0.71 | 9.12 | 0.39 |
| resins | 420 | 78.44 | 8.66 | 2.63 | 7.36 | 2.04 |
| asphaltenes | — | — | — | — | — | — |
| Cut III, total | 420 | 84.65 | 10.17 | 2.97 | 1.68 | 0.78 |
| saturates | 390 | 85.61 | 14.25 | — | — | — |
| aromatics | — | 84.35 | 10.40 | <0.01 | 4.44 | 0.90 |
| monoaromatics | — | 84.14 | 11.85 | <0.01 | 4.49 | <0.01 |
| di- + triaromatics | 370 | 86.23 | 9.54 | 0.08 | 1.75 | 1.73 |
| polyaromatics | — | — | — | 0.84 | 6.25 | 0.50 |
| resins | 950 | 79.95 | 8.56 | 2.50 | 3.58 | 0.93 |
| asphaltenes | [c] | 83.12 | 7.45 | 4.63 | 4.23 | 0.49 |

[a] From GLC data.
[b] From low voltage mass data.
[c] Will not dissolve in benzene.

**Table V.  Simulated Distillation**

| | Tar Sand Bitumen Saturates | Coal Liquids | | | |
|---|---|---|---|---|---|
| | | Big Horn | | Pitt Seam | |
| Fraction Off | | Total Distillates | Heavy Saturates | Total Distillates | Heavy Saturates |
| IBP (°F) | 402 | 136 | 460 | 132 | 440 |
| 1% off at | 460 | 170 | 480 | 170 | 457 |
| 5 | 504 | 197 | 520 | 195 | 501 |
| 10 | 557 | 223 | 539 | 215 | 532 |
| 20 | 601 | 273 | 569 | 247 | 563 |
| 30 | 629 | 326 | 598 | 289 | 592 |
| 40 | 670 | 367 | 622 | 335 | 633 |
| 50 | 706 | 394 | 647 | 371 | 667 |
| 60 | 743 | 420 | 690 | 404 | 710 |
| 70 | 786 | 450 | 728 | 431 | 760 |
| 80 | 828 | 483 | 769 | 470 | 798 |
| 90 | 870 | 532 | 814 | 520 | 858 |
| 95 | 891 | 564 | 849 | 551 | 897 |
| 99 | 940 | 623 | 894 | 602 | 984 |
| FBP | 954[a] | 696 | 917[a] | 664 | 1016[a] |

[a] GLC trace does not return to base line indicating that material boiling higher than the FBP is present.

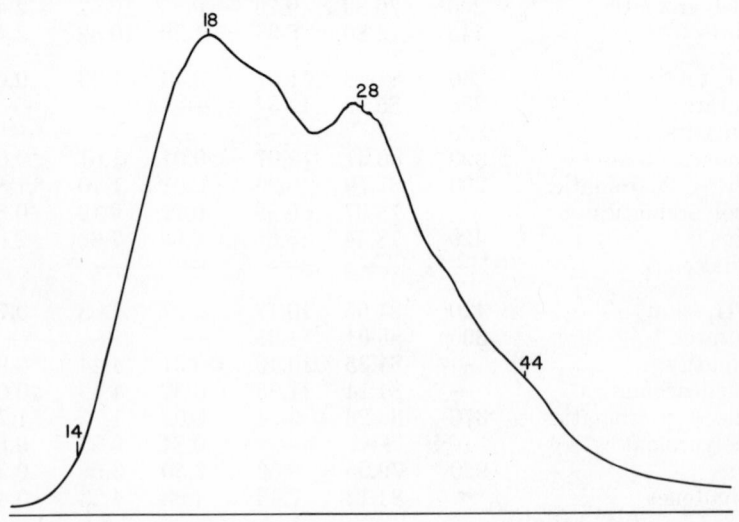

*Figure 2.  GLC of saturates from tar sands bitumen.  Numbers indicate n-paraffin standards.*

**Data (ASTM D2887)**

### Shale Oils

| Distillate From Cut I | Cut I | Cut II | Cut III |
|---|---|---|---|
| | | *Saturates* | |
| 118 | 433 | 481 | 491 |
| 146 | 443 | 517 | 545 |
| 174 | 463 | 562 | 621 |
| 204 | 480 | 576 | 659 |
| 235 | 498 | 600 | 697 |
| 261 | 515 | 613 | 749 |
| 289 | 531 | 624 | 779 |
| 312 | 545 | 630 | 806 |
| 342 | 557 | 641 | 824 |
| 361 | 565 | 657 | 848 |
| 391 | 572 | 682 | 879 |
| 435 | 585 | 703 | 918 |
| 465 | 594 | 736 | 956 |
| 516 | 615 | 797 | 1014 |
| 563 | 634 | 831 | 1094 |

**Table VI. Mass Spectrometric Analysis of Saturates (volume %)**

| | Tar Sand Bitumen | Big Horn Coal | Pitt Seam Coal | Cut I | Cut II | Cut III |
|---|---|---|---|---|---|---|
| | | *Coal Liquids* | | *Shale Oils* | | |
| Weight % of fraction in sample | 16.5 | 3.6 | 1.8 | 6.0 | 31.0 | 12.3 |
| Alkanes | 0.0 | 39.9 | 22.0 | 42.2 | 34.4 | 29.4 |
| Noncondensed cycloalkanes | 48.0 | 23.8 | 43.9 | 40.8 | 47.1 | 46.2 |
| Condensed cycloalkanes | 47.9 | 32.2 | 30.4 | 15.4 | 16.6 | 22.5 |
| 2 rings | 26.7 | 16.3 | 16.2 | 11.3 | 11.8 | 8.9 |
| 3 rings | 14.0 | 9.5 | 7.2 | 3.8 | 4.0 | 11.1 |
| 4 rings | 6.5 | 4.1 | 4.2 | 0.3 | 0.9 | 2.5 |
| 5 rings | 0.7 | 1.4 | 1.7 | 0.0 | 0.0 | 0.0 |
| 6 rings | 0.0 | 0.9 | 1.1 | 0.0 | 0.0 | 0.0 |
| Benzenes[a] | 4.0 | 3.8 | 3.3 | 1.6 | 1.8 | 1.8 |
| Naphthalenes[a] | 0.1 | 0.2 | 0.4 | 0.0 | 0.0 | 0.1 |

[a] Artifacts caused by the fragmentation of certain cycloparaffins.

**Table VII.    ¹H NMR Analysis**

| Sample Fraction | Mono- | Di- + Tri- | PNA |
|---|---|---|---|
| | *Tar Sands Bitumen* | | |
| Wt. % of fraction in sample | 7.0 | 30.1 | 10.7 |
| Monoaromatics (mole %) | 100.0 | 53.7 | 49.1 |
| Diaromatics (mole %) | 0.0 | 46.3 | 50.9 |
| Triaromatics (mole %) | 0.0 | <0.1 | <0.1 |
| Aromaticity ($f_A$) | 0.24 | 0.37 | 0.27 |
| Average molecular weight[a] | 339 | 284 | 404 |
| Alkyl substituents/molecule | 3.5 | 3.5 | 4.4 |
| Carbons/alkyl substituents | 5.3 | 3.7 | 4.9 |
| Aromatic rings/molecule | 1.0 | 1.5 | 1.5 |
| Naphthene rings/molecule | 1.4 | 0.9 | 1.5 |

[a] Does not account for O, N, S, etc.

The aromatic fraction accounts for almost half of the bitumen with the largest contribution made by the di- + triaromatics (Table II). The aromatic fractions were further characterized by a ¹H NMR spectroscopic technique. This method, developed for petroleum crudes and fractions, calculates from the NMR spectrum a set of average parameters used to describe an "average molecule." In this method, three assumptions are necessary which place constraints on its applicability.

1. No aromatic fused ring systems larger than three are present,

2. The resonances of the unsubstituted non-bridge aromatic ring carbon protons are sufficiently separated in the proton NMR spectrum so that the ratio of mono- to di- to triaromatics can be determined,

3. The number of substituent groups, on the average, is the same for mono-, di-, and triaromatics.

**Table VIII.    ¹H NMR Analysis**

| Sample Fraction | Mono- | Di- + Tri- | PNA |
|---|---|---|---|
| | *Cut I* | | |
| Wt. % of fraction in sample | 8.3 | 4.2 | 2.3 |
| Monoaromatics (mole %) | 100.0 | 38.2 | 81.0 |
| Diaromatics (mole %) | 0.0 | 61.8 | 19.0 |
| Triaromatics (mole %) | 0.0 | 0.0 | 0.0 |
| Aromaticity ($f_A$) | 0.20 | 0.61 | 0.32 |
| Average molecular weight[a] | 413 | 183 | 283 |
| Alkyl substituents/molecule | 4.4 | 3.3 | 4.6 |
| Carbons/alkyl substituents | 5.5 | 1.6 | 3.1 |
| Aromatic rings/molecule | 1.0 | 1.6 | 1.2 |
| Naphthene rings/molecule | 2.5 | 0.4 | 1.1 |

[a] Does not account for N, O, S, etc.

**of Aromatic Fractions**

| Big Horn Coal | | Pitt Seam Coal | |
|---|---|---|---|
| *Mono-* | *Di- + Tri-* | *Mono-* | *Di- + Tri-* |
| 9.0 | 42.0 | 5.9 | 26.7 |
| 100.0 | 16.4 | 100.0 | 15.6 |
| 0.0 | 69.7 | 0.0 | 63.5 |
| 0.0 | 13.8 | 0.0 | 21.0 |
| 0.39 | 0.76 | 0.37 | 0.73 |
| 204 | 169 | 218 | 181 |
| 3.4 | 1.8 | 4.0 | 2.1 |
| 2.7 | 1.7 | 2.6 | 1.8 |
| 1.0 | 2.0 | 1.0 | 2.1 |
| 1.3 | 0.3 | 1.5 | 0.4 |

The last two assumptions are probably as valid for synthetic liquid fuels as for petroleum. The first assumption is partially satisfied by the separation steps which provide three fractions two of which (the monoaromatics and the di- + triaromatics) are free of fused ring systems larger than three rings. The proton NMR analysis of the third fraction (the polyaromatics)—which contains four or more rings, condensed and noncondensed—is then only qualitative.

Figure 3 shows the spectra of the three aromatic subfractions and Table VII the results of the calculations. NMR showed the monoaromatic subfraction to be free from other aromatic types and to have many short alkyl substituents and at least one naphthene ring per molecule. The average molecular weight obtained by NMR agrees very well with that determined by VPO.

**of Aromatic Shale Oil Fractions**

| Cut II | | | Cut III | | |
|---|---|---|---|---|---|
| *Mono-* | *Di- + Tri-* | *PNA* | *Mono-* | *Di- + Tri-* | *PNA* |
| 8.0 | 13.4 | 5.6 | 5.3 | 9.3 | 1.3 |
| 100.0 | 52.8 | 84.3 | 100.0 | 55.0 | 56.4 |
| 0.0 | 47.2 | 15.7 | 0.0 | 35.5 | 28.5 |
| 0.0 | 0.0 | 0.0 | 0.0 | 9.5 | 15.1 |
| 0.26 | 0.51 | 0.29 | 0.18 | 0.44 | 0.44 |
| 316 | 204 | 311 | 448 | 248 | 248 |
| 4.0 | 3.6 | 5.2 | 5.4 | 3.9 | 5.5 |
| 4.3 | 2.1 | 3.2 | 5.0 | 2.7 | 1.9 |
| 1.0 | 1.5 | 1.2 | 1.0 | 1.5 | 1.6 |
| 1.5 | 0.5 | 1.2 | 2.2 | 0.7 | 0.3 |

The di- + triaromatic subfraction analyzes as 54% monoaromatics and 47% diaromatics by proton NMR. The average molecule also contains many short alkyl substituents, more than one aromatic ring, and one naphthene ring. The average molecular weight calculated from NMR is lower than that obtained by VPO. These data indicate that non-condensed di- and triaromatics are present in this subfraction. Compounds such as:

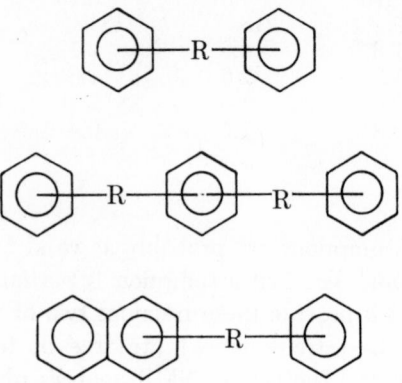

are di- and triaromatics and are all separated as such by the alumina column. Proton NMR, however, will see the first two as monoaromatics and the last one as a monoaromatic and a diaromatic, and all the calculations are affected accordingly. There is no way to circumvent this problem until the condensed and noncondensed aromatics can be separated from each other. It must be pointed out, however, that the presence of noncondensed systems could not have been detected by separation or spectroscopic techniques alone. Both must be used, and one must support the other.

The molecular weight of the polyaromatic fraction as calculated by NMR is well below that determined by VPO. As pointed out earlier the NMR analysis of this fraction can only be semiquantitative because tetra- and higher aromatic systems will be calculated as mono- and diaromatics and all the calculations will be affected accordingly. In our separation scheme all of the polar non-hydrocarbons are concentrated in the resin fractions. Only ethers and thioethers are included in the oil and are eventually concentrated in the di- + triaromatics and polyaromatics, as the data in Table III show. Also only half of the saturates are condensed cycloalkanes, mainly of two and three rings. These observations are indirect evidence that no significant amounts of large condensed systems are present and that at least part of the polyaromatic fraction consists of noncondensed mono-, di-, and triaromatic units.

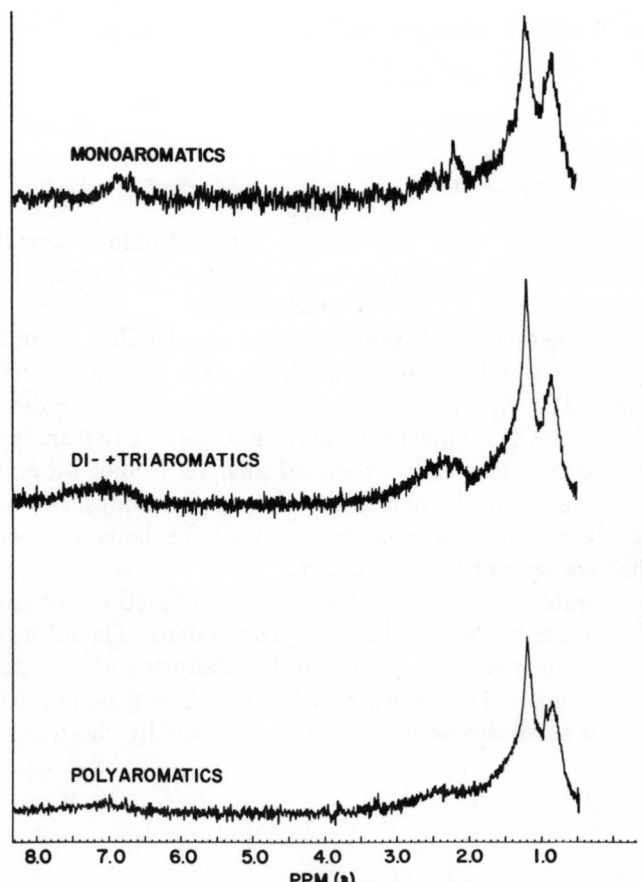

*Figure 3. $^1H$ NMR spectra of aromatic fractions from tar sands bitumen*

The resins and asphaltenes from tar sands and from the other synthetic fuels have not been analyzed beyond the extent shown by the tables. The resins can be fractionated and analyzed in more detail using methods developed for petroleum resins by Jewell (*46*) and McKay (*47*), but more extensive work is necessary to have definitive analysis methods for these materials.

**Coal Liquids.** The two coal liquids contain about the same amount of material boiling below 470°F, very little saturates, and substantial amounts of aromatics, mainly di- + triaromatics (Table II). The liquids from the Big Horn coal, however, contain more aromatics and less resins, asphaltenes, and benzene insolubles than the liquids from Pittsburgh Seam coal. This is not surprising considering the fact that higher rank coals are harder to hydrogenate and that the liquefaction process is believed to follow the path:

$$\text{Coal} \rightarrow \text{Benzene Insolubles} \rightarrow \text{Asphaltenes} \nearrow \begin{array}{c} \text{Resins} \\ \downarrow \\ \rightarrow \text{Aromatics} \end{array}$$

The distillates were fractionated into saturates, aromatics, and olefins by FIA as already mentioned. Attempts to further characterize the FIA fractions by GLC were not successful.

By GLC methods, it is possible to obtain detailed quantitative analyses of saturates up to $C_9$, of mono-olefins up to $C_7$, and of aromatics up to $C_{10}$ (48). The liquid chromatographic steps cannot handle materials boiling below $n\text{-}C_{12}$. Therefore, there is a gap in the analysis of the distillate fractions on which a detailed analysis cannot be readily, routinely, and inexpensively obtained. This gap includes all the $C_8$–$C_{12}$ saturates, all the $C_{11}$–$C_{12}$ aromatics, and all the heterocompounds and olefins that are present in this fraction.

The saturates boiling above 470°F were analyzed by MS and appear to contain alkanes but predominantly cycloparaffins. The alkanes appearing in these liquids may not be part of the coal network but may simply be embedded in it. The primary coal structure is generally believed to be formed of small aromatic units held together by short links, mainly

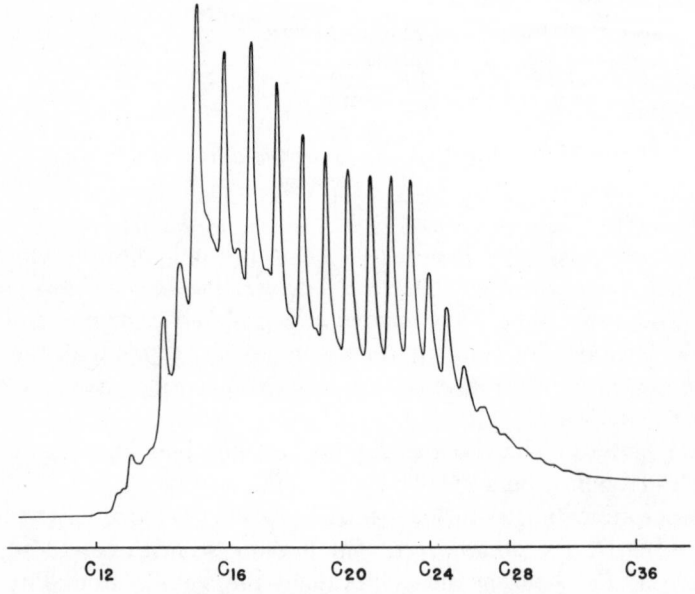

*Figure 4.   GLC of saturates from Big Horn coal liquids*

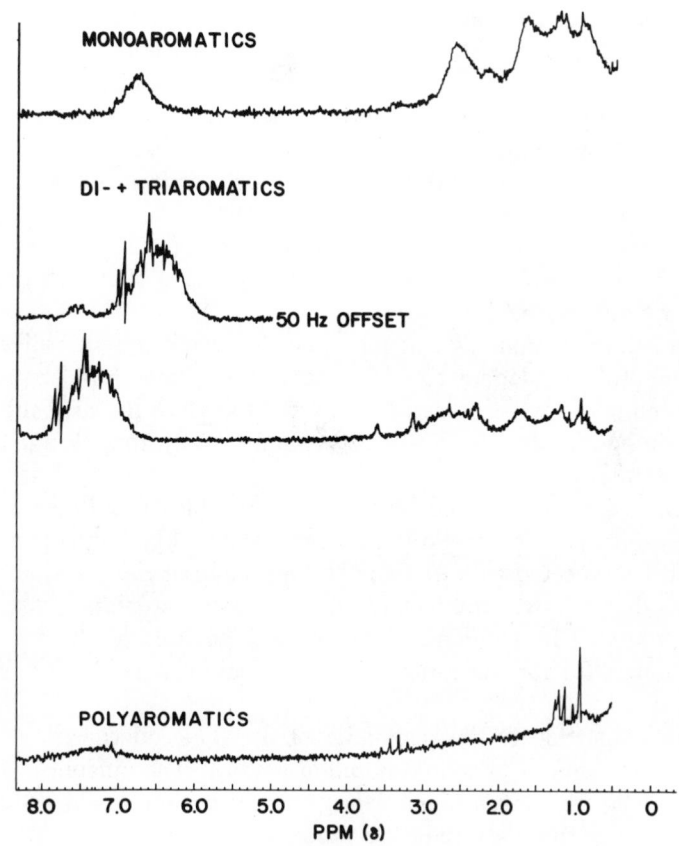

*Figure 5.*  *¹H NMR spectra of aromatic fractions from coal liquids*

methylene, ethylene, propylene, and ethers (*49–56*). However, the alkanes present in these fractions have a carbon range between 12 and 30, as shown by the chromatogram in Figure 4, and should not be the result of the coal decomposition. Vahrman *et al.* have shown that small molecules can be extracted from coal by non-destructive methods (*57–62*).

The cyclic saturates can also be entrained in the coal pores, but they can also be the result of the hydrogenation and liquefaction process. To establish the actual origin of these saturates, it would be necessary to carry out a considerable amount of work which is beyond the present scope.

The aromatic fractions were also examined by ¹H NMR, and examples of spectra are shown in Figure 5. The liquids are derived from coals of different ranks, and this is reflected in the size of each fraction as already pointed out. However, at least for these materials, the characteristics of corresponding fractions are very similar regardless of their origin. This is evident from the MS analysis of the saturates and is further shown by the analysis of the aromatic fractions. The NMR spectra of corresponding fractions from the two coal liquids are almost superimposable, and for this reason only one set of spectra is shown here, that for the Big Horn coal liquids. The similarity of the aromatics from the two coals is made quite clear by the NMR data in Table VII.

The two monoaromatic and di- + triaromatic fractions are practically indistinguishable from each other except for a slightly higher molecular weight of the fractions from the Pitt Seam coal liquids. The spectra of the polyaromatic fractions were too weak and unresolved, and no meaningful calculations could be made from them. Similar problems were encountered when it was attempted to analyze the asphaltenes by NMR. Methods have to be developed to analyze polyaromatic and asphaltene fractions.

**Shale Oils.** As indicated by the data in Table II, only cut I of the shale oils contains a considerable amount of material boiling below 470°F, as would be expected from the low boiling range of this fraction (Table V). This distillate was fractionated into saturates, aromatics, and olefins by preparative FIA techniques, but a GLC analysis of these fractions proved unfruitful for the same problems mentioned above for the coal liquids.

A large portion of the remainder of the shale oil consists of resins with smaller amounts of saturates and aromatics. The amount of asphaltenes is really insignificant, and a larger contribution would actually be surprising since these are retorting products.

The shale oils are rich in olefins; in our separation scheme free olefins are concentrated in the saturate fractions. The IR spectra of all three saturate fractions show the characteristic olefin bands at 6.1, 10.1, 10.35, and 11.0 $\mu$. However, no attempt has been made to characterize these olefins in detail, mainly because they are easily hydrogenated.

The mass spectrometric analyses of the saturate fractions are reported in Table VI. These fractions appear to be composed mostly of alkanes and noncondensed cycloalkanes with smaller amounts of condensed cycloalkanes, mainly two- and three-ring systems. However, because of the presence of olefins in these fractions the analyses are only semiquantitative. In fact, an olefin should make a contribution to the cycloalkane group type which has the same molecular weight. That is, a monoolefin will contribute to the cycloalkanes, a diolefin or a cyclic oleln will contribute to the bicycloalkanes, etc. However, to determine the extent of these contributions more analytical work is necessary.

No carbon number predominance was detected in the saturate fraction from cut I and cut II. The chromatogram of the saturates from

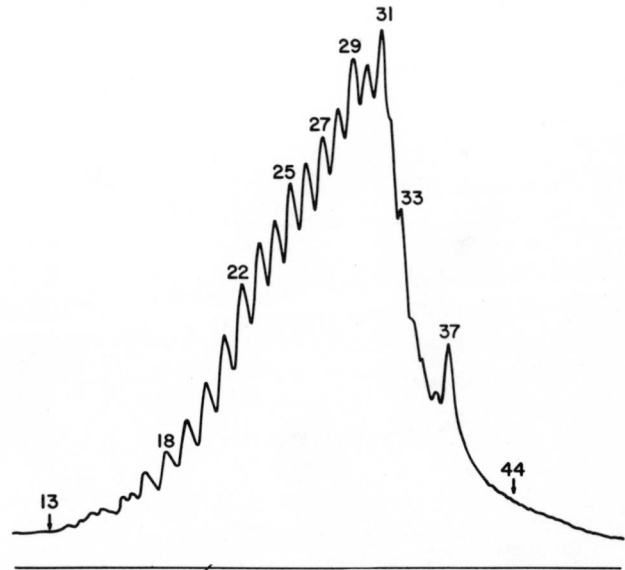

*Figure 6.   GLC of saturates from shale oils.   Numbers indicate n-paraffin standards.*

cut III is shown in Figure 6 and an odd-over-even carbon number predominance is evident for the higher carbon number region.

The aromatic subfractions have also been analyzed by proton NMR, and Figure 7 shows the spectra obtained with the fractions from cut II. The results of the calculations for all of the fractions are shown in Table VIII. The monoaromatics are free of other aromatic types, have a very low aromaticity, are highly substituted, and at least two of the substituents are naphthene rings. Anders *et al.* (*34*) and Gallegos (*35*) have isolated and identified many alkylbenzenes, alkyltetralines, dinaphthenebenzenes, and trinaphthenebenzenes from Green River shale oils.

The di- + triaromatics contain large amounts of noncondensed systems and a much higher aromaticity caused by a decrease in the number and size of the substituents. The apparent decrease in the average molecular weights (calculated) arises in part from the presence of noncondensed systems. As pointed out earlier, one molecule of 1,4-diphenylbutane, for example, will be detected by $^1$H NMR as two molecules of ethylbenzenes, and the calculation will be affected accordingly. An apparent decrease in the average molecular weight is then seen, and this decrease will be even more marked if the concentration of noncondensed triaromatics is significant. Anders and Robinson (*28*) have reported the evidence of large amounts of perhydrocarotenes, and Gallegos (*35*) has isolated and identified various phenyl (cyclohexyl) alkanes. By complete

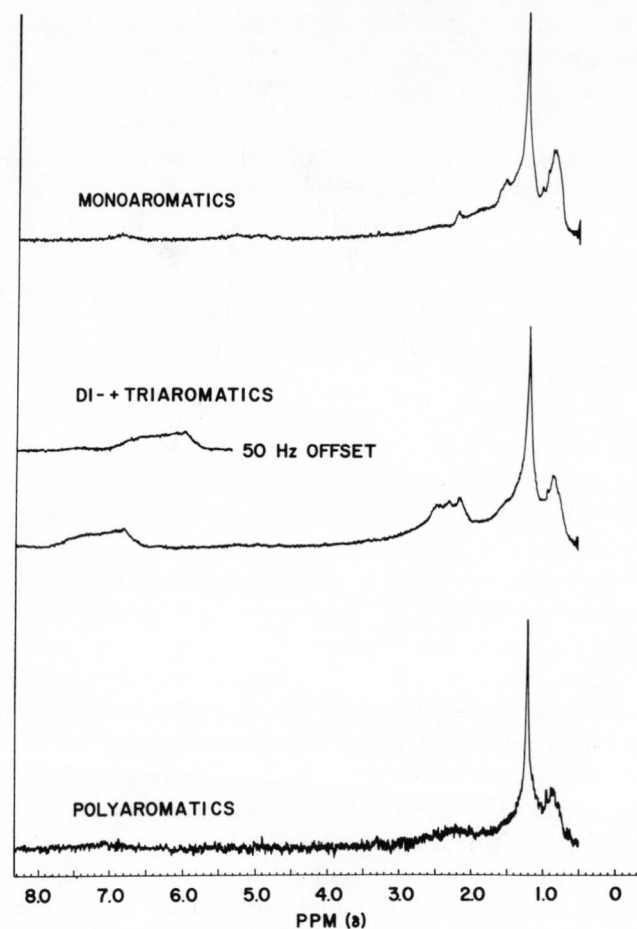

*Figure 7.*  ¹H NMR *spectra of aromatic fractions from shale oils*

dehydrogenation of these materials during maturation, noncondensed systems would be formed.

The ¹H NMR analyses of the polyaromatic fractions present the same problems already mentioned in the discussion of the corresponding fractions from other synthetics. While noncondensed systems are certainly present, no conclusion can be drawn from these data as to the true composition of these fractions.

**Comparison of Fuels.** The separation data of Table II show immediately some gross differences and similarities among the various fuels analyzed. Both coal liquids have a considerable amount of low boiling material which is made up of saturates and aromatics in about equal concentration. Similarly, the shale oil contains low boiling material while the tar sand bitumen does not.

Of the material boiling above 470°F, the tar sand bitumen contains more saturates than the other fuels. As pointed out above, these saturates are unique in that they do not contain free alkanes while the saturates from the other fuels contain 22–40% alkanes as determined by MS (Table VI).

Only the Big Horn coal liquids have more condensed cycloalkanes than noncondensed cycloalkanes. The tar sands have an equal distribution of the two types of saturates, and the remaining fuels have a higher concentration of noncondensed cycloalkanes. The effect of the olefins on the mass analysis of the saturates from the shale oils must, however, be kept in mind in making this comparison.

The fact that only the coal liquids show condensed systems with up to six rings and that the Pitt Seam coal liquids (products from a higher rank coal) have a higher concentration of these condensed systems is also worth pointing out. However, the amount of the total saturates in the coal liquids is insignificant from a production and refining point of view.

The aromatic content of the fuels ranges from about 15 to 58%, and in every case the largest contribution is made by the di- + triaromatics. In addition to the differences in the quantity of the aromatic fractions, as shown by the data in Table II, the characteristics of each fraction vary depending on its origin. The variations are shown pictorially by the NMR spectra and more tangibly by the NMR data listed in Tables VII and VIII. The tar sand bitumen fractions have, in general, a lower aromaticity than the shale oil and coal liquid fractions. This is caused by a higher number and the larger size of the substituents. The coal liquids have the highest concentration of condensed di- + triaromatics and the tar sand bitumen the lowest. Conversely, the amount of noncondensed systems is highest in the tar sand bitumen and lowest in the coal liquids.

Perhaps, where these fuels differ most is in the amount of resins, asphaltenes, and benzene insolubles. The shale oils contain a very large amount of resins from 12 to 69%, or approximately 50% by weight on a total shale oil basis. The tar sand bitumen contains only half as much resins, and the coal liquids contain much less. On the other hand, the asphaltenes are significant in the tar sand bitumen and in the Pitt Seam coal liquids. These latter materials are the only ones to contain a significant amount of benzene insolubles.

### Conclusion

Coal liquids and other synthetic liquid fuels can be analyzed by modifying the methods normally used for petroleum crudes and products.

These analysis methods are relatively fast, require only a few grams of sample, provide discrete fractions which can be characterized in as much detail as desired with available techniques, and provide comparative compositional profiles for fuels from various sources, both natural and synthetic. The methods have limitations, as indicated throughout the paper. However, as new procedures are developed to take care of these limitations, they can readily be incorporated.

## Acknowledgments

The authors wish to thank A. V. Vayda for aid in preparing the coal liquids; W. E. Magison for help in obtaining the NMR data; W. Hubis and H. P. Malone for helpful discussions; The Oil Shale Corp. and Sun Oil Co. for providing the samples of shale oil and tar sand bitumen, respectively, and for their permission to publish the data obtained on them.

## Literature Cited

1. Camp, F. W., "Tar Sands," *Encycl. Chem. Technol.*, 2nd Ed. (1969) **19**, 682.
2. Carrigy, M. A., Ed., "Athabasca Oil Sands: A Collection of Papers Presented to K. A. Clark on His 75th Birthday," Research Council of Alberta, Edmonton, Alberta, Canada, Oct., 1963.
3. Boyd, M. L., Montgomery, D. S., "Composition of Athabasca Bitumen Fractions as Determined by Structural-Group Analysis Methods," "Athabasca Oil Sands: A Collection of Papers Presented to K. A. Clark on His 75th Birthday," Research Council of Alberta, Alberta, Canada, 1963.
4. Speight, J. G., *Fuel, London* (1970) **49**, 76.
5. *Ibid.* (1971) **50**, 102.
6. Moschopedis, S. E., Speight, J. G., *Fuel, London* (1971) **50**, 34.
7. Moschopedis, S. E., *Fuel, London* (1971) **50**, 211.
8. Special Section, "Shale Oil—The Problems and Prospects," *Oil Gas J.* (March 9, 1967), 65.
9. Gustafson, R. E., "Shale Oil," *Encycl. Chem. Technol.*, 2nd Ed. (1969) **18**, 1.
10. Robinson, W. E., Dinneen, G. U., *Proc. World Pet. Congr., 7th* (1967) **3**, 669.
11. Cane R. G., *Proc. World Pet. Congr., 7th* (1967) **3**, 681.
12. Ball, J. S. *et al.*, *Ind. Eng. Chem.* (1949) **41**, 581.
13. Dinneen, G. U. *et al.*, *Ind. Eng. Chem.* (1952) **44**, 2632.
14. Cady, W. E. *et al.*, *Ind. Eng. Chem.* (1952) **44**, 2636.
15. Dinneen, G. U. *et al.*, *Ind. Eng. Chem.* (1952) **44**, 2647.
16. Dinney, I. W. *et al.*, *Anal. Chem.* (1952) **24**, 1749.
17. Van Meter, R. A. *et al.*, *Anal. Chem.* (1952) **24**, 1758.
18. Dinneen, G. U. *et al.*, *Anal. Chem.* (1955) **27**, 185.
19. *Ibid.* (1958) **30**, 2026.
20. Heady, H. H. *et al.*, *Bur. Mines Rep. Invest.* (1960) **5662**.
21. Morandi, J. R., Jensen, H. B., *J. Chem. Eng. Data* (1966) **11**, 81.
22. Jensen, H. B. *et al.*, *Am. Chem. Soc. Div. Fuel Chem, Preprints* (1968) **12** (1), F98.
23. *Ibid.* (1971) **15** (1), 113.

24. Robinson, W. E. *et al.*, *Geochim. Cosmochim. Acta* (1965) **29**, 249.
25. Gelpi, E. *et al.*, *Anal. Chem.* (1971) **43**, 864.
26. Gelpi, E. *et al.*, *J. Chromatogr. Sci.* (1971) **9**, 147.
27. Gallegos, E. J., *Anal. Chem.* (1971) **43**, 1151.
28. Anders, D. E., Robinson, W. E., *Geochim. Cosmochim. Acta* (1971) **35**, 661.
29. Simoneit, B. R. *et al.*, *Chem. Geol.* (1971) **7**, 123.
30. Douglas, A. G. *et al.*, *Tetrahedron* (1971) **27**, 1071.
31. Djuricic, M. *et al.*, *Geochim. Cosmochim. Acta* (1971) **35**, 1201.
32. Haug, P. *et al.*, *Chem. Geol.* (1971) **7**, 213.
33. Maxwell, J. R. *et al.*, *Geochim. Cosmochim. Acta* (1973) **37**, 297.
34. Anders, D. E. *et al.*, *Geochim. Cosmochim. Acta* (1973) **37**, 1213.
35. Gallegos, E. J., *Anal. Chem.* (1973) **45**, 1399.
36. Lowry, H. H.,*Chem. Coal Util., Suppl. Vol.* (1963).
37. Van Krevelen, D. W., "Coal," Elsevier, New York, 1961.
38. Seglin, L., Eddinger, R. T., *Encycl. Chem. Technol.*, 2nd Ed., Suppl. Vol. (1971), 178.
39. Speight, J. G., *Appl. Spectrosc. Rev.* (1972) **5**, 211.
40. Jewell, D. M. *et al.*, *Am. Chem. Soc. Div. Petrol. Chem.*, Preprints (1972) **17** (4), F81.
41. O'Connor, J. G. *et al.*, *Anal. Chem.* (1962) **34**, 82.
42. Jewell, D. M. *et al.*, *Anal. Chem.* (1972) **44**, 2318.
43. Aluise, V. A. *et al.*, *Anal. Chem.* (1947) **19**, 347.
44. Hood, A., O'Neal, M. J., *Adv. Mass Spectrom.* (1959) 175.
45. Clutter, D. R. *et al.*, *Anal. Chem.* (1972) **44**, 1395.
46. Jewell, D. M. *et al.*, *Anal. Chem.* (1972) **44**, 1391.
47. McKay, J. F., Latham, D. R., *Anal. Chem.* (1972) **44**, 2132.
48. Mayer, T. J. *et al.*, *Am. Chem. Soc. Div. Petrol. Chem.*, Preprints (1972) **17** (4), F14.
49. Heredy, L. A. *et al.*, ADVAN. CHEM. SER. (1966) **55**, 493.
50. Heredy, L. A. *et al.*, *Fuel, London* (1965) **44**, 125.
51. Chatterjee, A. K., Mazundar, B. K., *Fuel, London* (1965) **47**, 93.
52. Ouchi, K., Brooks, J. D., *Fuel, London* (1967) **46**, 367.
53. Bartle, K. D., Smith, J. A., *Fuel, London* (1965) **44**, 109.
54. Lawson, G. J., Purdie, J. W., *Fuel, London* (1966) **45**, 115.
55. Davies, L., Lawson, G. J., *Fuel, London* (1967) **46**, 95.
56. Chakrabarthy, S. K., Kretschmen, H. O., *Fuel, London* (1972) **51**, 160.
57. Vahrman, Mark, *Fuel, London* (1970) **49**, 5.
58. Rahman, M., Vahrman, M., *Fuel, London* (1971) **50**, 318.
59. Palmer, T. J., Vahrman, M., *Fuel, London* (1972) **51**, 14.
60. *Ibid.*, 22.
61. Vahrman, M., Watts, R. H., *Fuel, London* (1972) **51**, 130.
62. *Ibid.*, 235.

RECEIVED December 16, 1974.

# 4

# Rate of Dissolution of Carbonate Mineral Matrix in Oil Shale by Dilute Acids

M. MOUSSAVI, D. K. YOUNG, and T. F. YEN

Department of Chemical Engineering, University of Southern California, Los Angeles, Calif. 90007

*To further understand the inner nature of oil shale matrix, samples of different particle sizes were studied. Oil shale of Green River formation was dissolved with dilute acids at pH 1–1.5 and 0.7–0.8, the weight loss of original sample was measured, and the magnesium ion was determined by atomic absorption spectrometry. The rate of leaching of mineral matrix did not follow the uniform and homogeneous spherical models. Kinetics of the second order reactions was observed. The range of particle size in which the experiments were conducted showed the possibility of total breakdown of mineral matrix in the oil shale by a bioleaching process.*

Almost all of the pilot plant and commercial scale shale oil recovery systems are based, thus far, on applying heat by thermal degradation methods. The low recovery efficiency and numerous environmental impacts associated with these methods are the main drawbacks.

In very few cases, for research purposes, certain chemical methods such as oxidation, hydrogenation, or application of some other reagents which could attack the sensitive sites of a kerogen subunit have been applied. However, research workers in this field are inclined to attack the kerogenic material by thermal degradation because retorting is the only common practice.

But it seems that the kerogen could be released also by attacking the inorganic portion of the oil shale system. This would result in higher efficiency and less environmental impacts. For example, efforts undertaken by this group using biological activities such as bioleaching or enzymatic techniques indicate that kerogen could be successfully released from the minerals matrix by a non-retorting method.

Because of the combination of physical, mechanical, and architectural properties of the mineral core, which is in tight contact with the organic phase, it is difficult for attacking agents to permeate the whole system. This may be the major reason why investigators have become discouraged with this approach.

Application of surface active agents has demonstrated that kerogen forms a protective coating around the mineral matrix. This coating impedes the contacting of mineral with leachable liquid (1) and may be the primary cause of the impermeability and stress properties of the oil shale (2).

It is highly probable that metal ions are involved in the structural conformation of the organic matter as in its algal (3) originators. However, the presence of strong bonds between organic and inorganic phases is yet to be proved (4, 5). This study reveals certain inner nature of the oil shale aggregate by analyzing the kinetic data obtained from disaggregation of its carbonate part.

The kerogenic material present in the Green River shale which is of Eocene Age is bound to a mineral matrix composed of carbonates, quartz, clays, and other minor minerals.   The isolation of this organic material can, therefore, be carried out in two ways.

1.   The complete disintegration and dissolution of the mineral material.   This method frees the largely intact kerogen.  Concentrated hydrofluoric acid is the most widely used reagent in this class (6, 7) since it is still the only reagent that can effectively solubilize the silicate mineral (clay and quartz) that is resistant to most chemical treatments.

2.   The partial cracking of the kerogen into smaller and more soluble components while leaving the majority of the mineral matrix intact. Processes such as oxidation (8), pyrolysis (9), and hydrogenation (10), all belong to this category.

Methods to isolate the kerogen, preferably in the form of a more directly usable fuel, from raw shale with a minimal amount of pretreatment and handling are of obvious economic interest.  Knowledge of the physical structure of the intact raw shale would help in the design of such methods.  In particular, increasing the porosity of the raw shale without causing extensive disaggregation of the mineral matrix would facilitate the movement of materials through the mineral matrix.

Scanning electron micrographs (SEM) of bioleached shale, which has been leached with the acid produced by sulfur-oxidizing bacteria, have revealed a pitted, spongy-appearing surface texture. Bioleaching removes primarily the carbonate minerals, such as dolomite and calcite, which are apparently deposited in pits throughout the rest of the mineral matrix. The removal of the carbonate would be expected to increase the porosity of the raw shale. Since the results of SEM reveal only the surface effect of bioleaching, further processes are necessary to study the overall effects of bioleaching on the structure inside the raw shale. In order to understand the internal structure changes we have measured the kinetics of this dissolution process using ground shales of different particle size.

## Experimental

The samples were treated with an acidic reagent and the results were obtained by two different methods. (1) determination of the weight of undissolved material and (2) determination of the weight of the dissolved material due to the treatment. All shale samples used were collected from the Mahogany Ledge of the Green River formation. The shale rock was crushed and sized into three ranges—42/60, 60/100, and 150/200 of Tyler scale. Since the effect of bioleaching is essentially that of dilute acid reaction with the carbonate material, AR grade hydrochloric acid was used instead of the acid produced by the sulfur-oxidizing bacteria for better monitoring accuracy. The dissolution rate of the carbonate mineral was estimated by direct measurement of the weight loss using atomic absorption spectrophotometry.

**Direct Weight Loss Measurement** (method 1). In this method several containers were charged, each with 0.3 g of raw shale of known particle size and 50 ml of an HCl–KCl solution with ionic strength of about 0.1 molar ions/l. and pH of 1.28. The reaction was allowed to progress; after a specific period the shale from one of the containers was quickly separated from the acid by vacuum filtration and washed thoroughly with deionized water. The shale was then dried at 105°C for at least 2 hr before weighing. The same process was repeated for a pre-weighted container, and the data were adjusted.

**Dissolved-away Materials Measurement** (method 2). Calcium and magnesium are among the ions appearing in the solution from acid leaching of the dolomitic and calcitic parts of the shale minerals. However, magnesium is distributed more than twice as much as calcium when shale is treated by alkaline fusion. In this method, a 2-g sample of appropriate mesh size was mixed with 200 ml of dilute hydrochloric acid in distilled water at pH 0.8. Samples were taken after each given period, and the concentration of $Mg^{2+}$ was measured by an atomic absorption spectrophotometer at 285 nm. The above process was repeated for all different mesh sizes used in the first method. The hydrogen ion concentration was found to be almost constant throughout the experiments (pH 0.8–0.9).

## Results and Discussion

The results of the weight loss for 42/60, 60/100, and 150/200 mesh size shales (Method 1) are presented in Figure 1. The results shown in Figure 1 can be fitted, empirically, to the following equation:

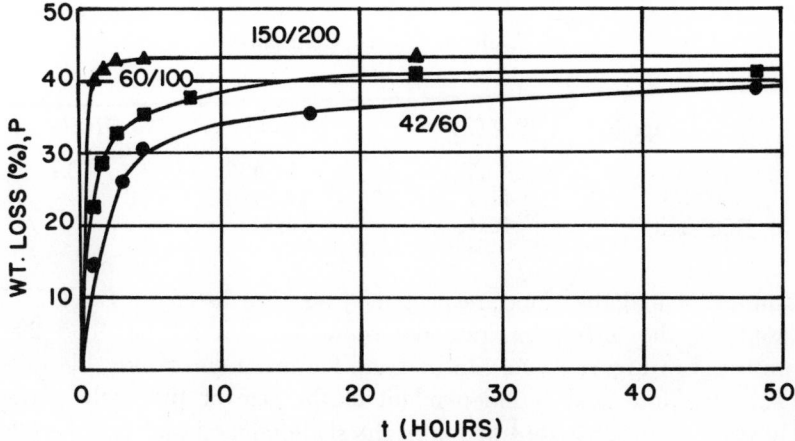

Figure 1. Rate of weight loss in dilute acid

$$\frac{P}{P_o - P} = Kt \qquad (1)$$

where $P$ is the percent weight loss at time, $t$; $P_o$ is the maximum percent weight loss; and $K$ is a constant. By rearranging this equation into the form:

$$P = P_o - P/Kt \qquad (2)$$

and plotting $P$ vs. $P/t$, straight lines are obtained such as in Figure 2. The values of $P_o$ and $K$ for each mesh size can be estimated from the

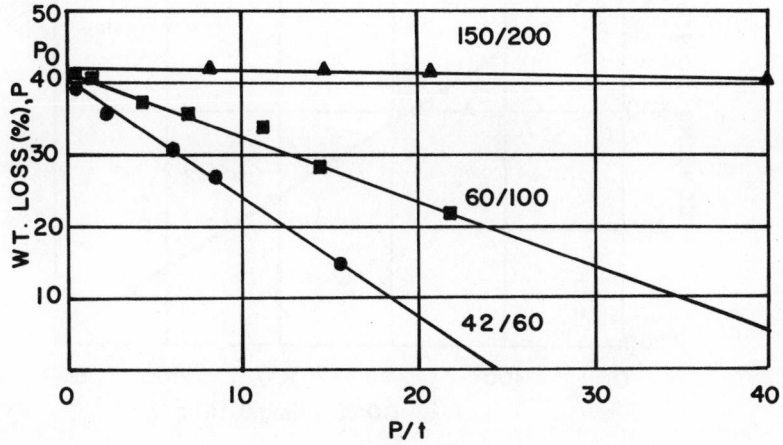

Figure 2. Rate constant determination by the equation $P = P_o - P/Kt$ (method 1)

### Table I.   Rate Constants

| | | $K(hr^{-1})$ | |
| Mesh Range | $P_o$ (%) | Method 1 | Method 2 |
|:---:|:---:|:---:|:---:|
| 42/60 | 40.7 | 0.59 | 0.7 |
| 60/100 | 41.0 | 1.10 | |
| 150/200 | 42.8 | 13.30 | |

Y-intercept and the slope, respectively, and are listed in Table I. The results of the dissolution rate determination method (method 2) are presented in Figure 3. As can be seen, the maximum removable mineral ($P_o$) by dilute acid is independent of the size of the shale particles. However, the carbonate fraction in the shale mineral matrix is very close to this figure. This could mean that the accessibility of the leaching agent to the leachable materials in shale is complete in the size ranges studied in this experiment—but at different rates. This could also indicate that the carbonate deposit sites are not isolated but can, perhaps, be thought of as interconnected by minerals built of the dilute acid-resistant minerals.

A comparison between the $K$ values obtained by the above two methods shows that for the mesh range of 40/60, the results are reasonably close ($K_1 = 0.59$, $K_2 = 0.7$ hr$^{-1}$). But for the finer particles such as

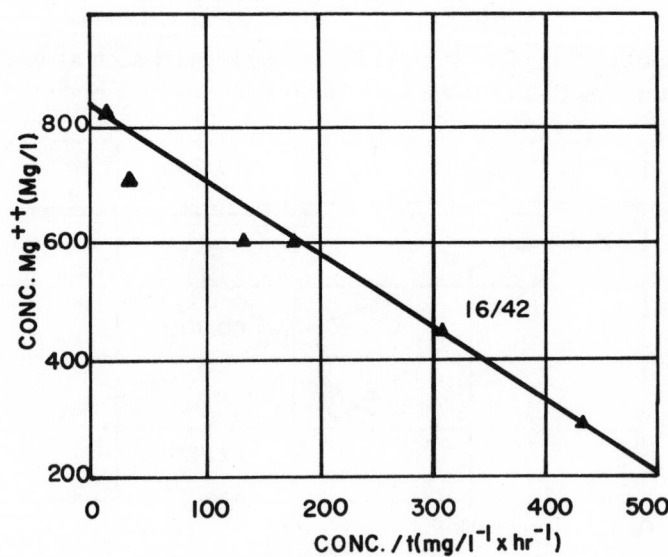

Figure 3.   Rate constant determination by the equation $P = P_o - P/Kt$ (method 2)

150/200, $K$'s do not match with each other. This is partly the result of the effects of the specific surface which was different in the two methods. However, the mechanisms of the dissolution kinetics seem to be identical. The reaction rate of the acid with carbonate mineral would be controlled by diffusion of the reactant into and the products out of the pores. Therefore, the availability of only the contact surface is not adequate. The type of surface in terms of relevant diffusion model and the closest theory to that model, such as film, penetration, or any other, should also be specified.

If the reaction is controlled by surface availability, then the rate of carbonate removal is, for spherical particles proportional to 2/3 powers of the unreacted mineral:

$$\frac{dP}{dt} = A \, (P_o - P)^{2/3} \qquad A = \text{constant} \tag{3}$$

The figure 2/3 is the dimension of the specific surfaces such as the case of the sublimation rate of mothball in the air. However, the presence of non-carbonate minerals and organic structures (kerogen matrix) which separate the carbonate mineral into isolated sites connected by channels would alter the rate so that:

$$\frac{dt}{dP} = A \, (P_o - P)^n \qquad n \neq 2/3 \tag{4}$$

The empirical Equation 1 is derived by integrating with limits from 0 to $P$:

$$\frac{dP}{dt} = (K/P_o) \, (P - P_o)^2 \qquad n = 2 \tag{5}$$

Because of the obvious complexity of the controlling factors that may be involved, we shall not attempt at present to estimate their functional magnitude. The existence of interconnecting channels was postulated from scanning electron micrographs of bioleached shale surface (11) and is supported by evidence from the present study. The increase of porosity by removing the carbonate mineral with dilute acid would presumably improve the permeability of certain chemical compounds into and out of the remaining shale structure.

*Summary*

The results of this work indicate that the process of disaggregation of the minerals matrix can be 80–100% completed within a few hours in the range of mesh sizes which was used in this experiment.

They also indicate the feasibility of *in situ* leaching and eventual recovery in conjunction with an oxidation method. This may be economically compatible with conventional thermal degradation and retorting methods.

## Acknowledgment

This work was supported by NSF Grant No. GI-35683, AER-74-23797, and A.G.A. BR-48-12.

## Literature Cited

1. Tackett, J. E., Jr., U.S. Patent **3,503,705**, C. 3131/70, f3/4/68, Appl. 709,909.
2. Tisot, P. R., Sohns, H. W., *Amer. Chem. Soc. Div. Pet. Chem., Preprint* 14, (3), B94-B104.
3. Cane, R. F., *Geochim. Cosmochim. Acta* (Feb. 1969) 33, (1), 257–265.
4. Thomas, R. D., Lorenz, P. B., *U.S. Bur. Mines, Rep. Invest.* (April 1970) 7378.
5. Tisot, P. R., *J. Chem. Eng. Data* (1962) 1, 405–410.
6. Robinson, W. E., "Organic Geochemistry," (G. Ellington and M. T. J. Murphy, Eds.), Chap. 6, pp. 181–195, Springer-Verlag, Berlin, 1969.
7. Saxby, J. D., *Chem. Geol.* (1970) 6, 173–184.
8. Robinson, W. E., *et al.*, *U.S. Bur. Mines Rep. Invest.* (1963) **6166**.
9. Dinneen, G. U., *Chem. Eng. Prog. Symp. Ser.* (1965) **61** (54), 42–47.
10. Shultz, E. B., Jr., *Chem. Eng. Prog. Symp. Ser.* (1965) **61** (54), 48–59.
11. Meyer, W. C., Yen, T. F., "Effects of Bioleaching on Oil Shale," *Amer. Chem. Soc., Div. Fuel Chem.; Preprints* (1974) **19** (2), 34.

RECEIVED September 12, 1975.

# Hydrogasification of Oil Shale

SANFORD A. WEIL, HARLAN L. FELDKIRCHNER, and
PAUL B. TARMAN

Institute of Gas Technology, Chicago, Ill. 60616

*Colorado oil shale was hydrogasified at temperatures up to 1400°F and pressures up to 500 psia. Both laboratory thermobalance tests and bench-scale moving-bed reactor tests were conducted. The thermobalance tests showed that at elevated hydrogen partial pressures with controlled slow shale heating, over 95% of the shale's organic carbon could be recovered. These results were verified in the bench-scale tests in which countercurrent hydrogen–shale contacting achieved organic carbon recoveries as high as 95%. Mineral carbonate decomposition was suppressed significantly by adding carbon dioxide to the feed gas.*

In the early 1960's, the Institute of Gas Technology (IGT) carried out an extensive research program for the American Gas Association (A.G.A.) on the direct hydrogasification of, primarily, Colorado oil shale (1) using laboratory and bench-scale equipment. In the bench-scale tests, using co-current hydrogen–shale contacting, about 65% of the organic carbon in the oil shale was gasified. About 15% was converted to light aromatic liquids, and about 20% remained in the spent shale. The major variable controlling the conversion of organic matter to gas was the hydrogen/shale ratio.

Economic studies had indicated that the cost of pipeline gas from oil shale was promising; however, there were two major disadvantages in co-current operation:

1. Approximately 20% of the available energy in the raw shale was not recovered, and it was questionable whether the residual organic matter could be economically recovered as process fuel.

2. Most of the heat used in preheating the shale to reaction temperature would be lost as sensible heat in the hot spent shale, which would be discharged at 1100°–1300°F.

This study was initiated, therefore, to investigate the technical and economic feasibility of producing synthetic pipeline gas from oil shale by hydrogasification, using countercurrent gas–solids contacting and excess hydrogen. The use of excess hydrogen with controlled, counter-current shale heating and cooling was expected to increase organic carbon removal from the shale and improve heat recovery, thus improving overall process efficiency compared with previous processes. The use of excess hydrogen would require hydrogen recycle but would probably permit operation at lower pressures and thus reduce plant capital equipment costs (relative to processes using co-current gas–solids contacting and near-stoichiometric hydrogen/shale ratios). Preliminary economic studies based on 20–25 gal/ton Fischer assay oil shale indicated that raw material costs would be much greater than reactor costs; thus, it would be desirable to convert as much of the organic matter content of the shale as possible. This would also be desirable in view of long range energy conservation.

Initially, we carried out material and energy balances, heat transfer, and other process design calculations to determine the general areas for the most efficient operation. Next, we conducted a series of tests using a thermobalance to determine the effects of temperature, pressure, and hydrogen level on kerogen conversion. Ways to minimize undesirable mineral carbonate decomposition reactions were also studied. Finally, a series of tests was conducted in a 4-in. diameter reactor to determine the effects of process variables on the yields of gaseous, liquid, and solid products and their rates of production from oil shale kerogen. In this paper, we present test results relating process variables to product yields and product properties.

### Thermobalance Studies

**Experimental.** The test program to study the effects of temperature, hydrogen pressure, particle size, and heat-up rate on the rate and extent of kerogen removal from the shale was carried out with a thermobalance, as diagramed in Figure 1. The design and operation of the equipment have been described in an earlier publication (2).

In most of the runs, the shale sample (as pebbles) was contained in a stainless steel, wire screen basket, 1/4–3/8 in. in diameter and 3 in. high. The reactor section was brought to the desired initial temperature with a stabilized gas flow stream at about 10 standard cu ft/hr. The sample was lowered into the reactor, and the power to the heating elements was adjusted to achieve the desired heat-up rate. The heat-up rates used included slow (about 15°F/min), fast (about 35°F/min), and very rapid, in which case the sample was lowered quickly into a preheated reactor. In the last case, we estimate that about 40 sec were required for the sample to attain reactor temperature. After a prescribed time or upon reaching a prescribed temperature, the sample was raised into a cool region above the reactor, effectively stopping further reaction.

The sample was weighed before and after each run, and the residue was analyzed for carbon, carbonate, hydrogen, and ash. During the run,

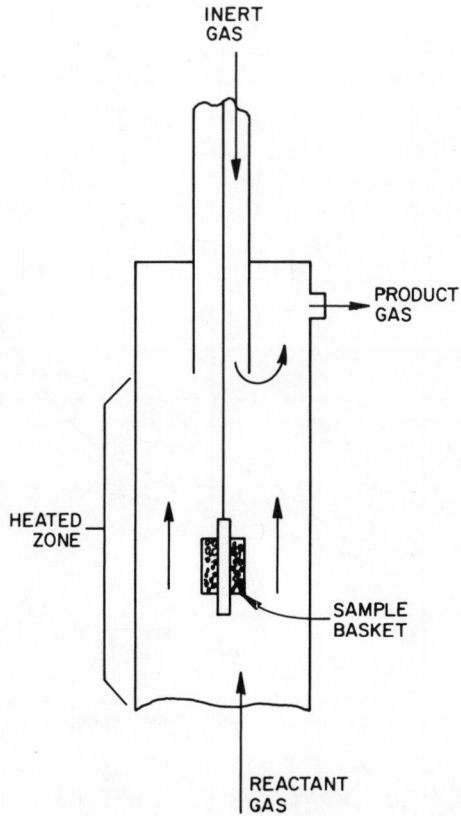

*Figure 1.    Thermobalance reactor*

the sample weight and temperatures (using four thermocouples, which surrounded the sample) were continuously recorded.

After selection, the shale rocks were crushed, sieved, and divided . into small samples by riffling. The shale samples weighed 2.5–3 g and were −6+10 U.S. Standard sieve size. The average composition of random samples chosen for chemical analysis is given in Table I.

**Results of Rich-Shale Studies—Total Conversion.** The weight loss *vs.* time curves and the shale residue compositions were determined for

### Table I. Shale Compositions

| Composition | Rich Shale, 52 gal/ton (wt %) | Lean Shale, 11 gal/ton (wt %) |
|---|---|---|
| Organic carbon | 21.06 | 5.40 |
| Hydrogen | 2.84 | 0.99 |
| Carbon dioxide | 12.54 | 18.17 |
| Sulfur | 0.83 | 0.45 |
| Calcium | 6.3 | 8.8 |
| Magnesium | 3.3 | 4.8 |
| Ash | 60.0 | 74.0 |

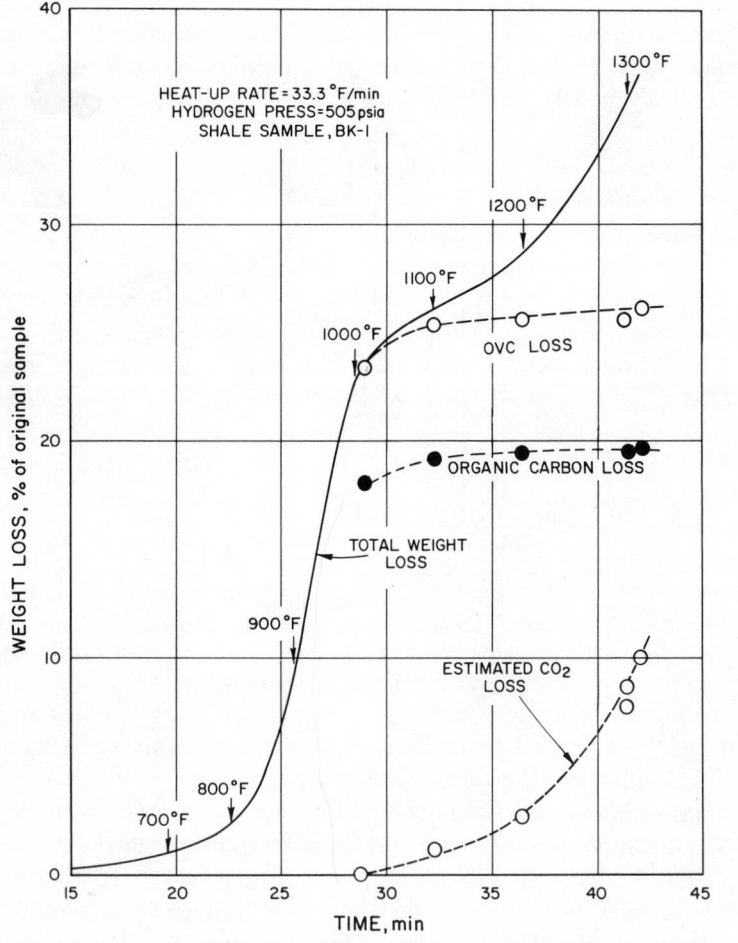

*Figure 2. Typical weight loss curve for rich shale*

a variety and range of conditions. The major variables were hydrogen pressure, gas composition, heat-up rate, and initial and maximum temperatures.

A typical time–temperature–weight loss relation is shown in Figure 2. Characteristics common to all heat-up rate runs are the onset of significant weight loss at 700°F and a rapid rate of weight loss between 800° and 1000°F, followed by a much slower rate until about 1100°F, when it increases again significantly. The thermobalance measures only weight changes, so the curve represents weight loss as a result of carbon dioxide as well as of kerogen removal. In any given run, only the final amounts of each component are known. Other runs were made at the same operating conditions and temperature histories, except that the samples were withdrawn from the reactor at different temperatures. From their residue analyses, points were added to Figure 2 to show the estimated weight loss from other volatile components besides carbon dioxide (OVC loss) and that from carbon dioxide loss. These results show that there is a sharp drop in the rate of kerogen removal above about 1000°F. Furthermore, although there is still organic carbon left, not much more weight is lost beyond 1200°F. This behavior appears to apply to all of the runs made.

Under almost any conditions, at least 87% of the organic carbon was recovered in a hydrogen atmosphere. The maximum recovery in a helium atmosphere was 77%.

On the basis of runs made with a variety of temperature histories, the following qualitative characteristics were true in a hydrogen atmosphere.

1. Direct exposure (hence, very rapid heat-up) to temperatures of 1300°F or higher leaves about 13% of the original carbon in the residue regardless of exposure time.

2. Soaking the shale sample for 1 hr or more at 700° or 800°F (or very slow heat-up) improves kerogen conversion at higher temperatures.

3. Soaking at low temperatures without heating above 1000°F results in as much kerogen recovery as direct exposure to high temperatures. (Thirteen percent of the original carbon remains.)

4. Higher hydrogen pressures resulted in lower residual organic carbon.

Figure 3 shows the residual organic carbon for the range of hydrogen pressures and heat-up rates studied. In all of these runs, the final temperature was 1300°F. Apparently kerogen recovery above 1000°F is extremely pressure sensitive.

Carbonate decomposition does not become significant until 1000°F and, as would be expected, is greater for longer periods at temperatures above 1000°F. There appears to be a qualitative correspondence be-

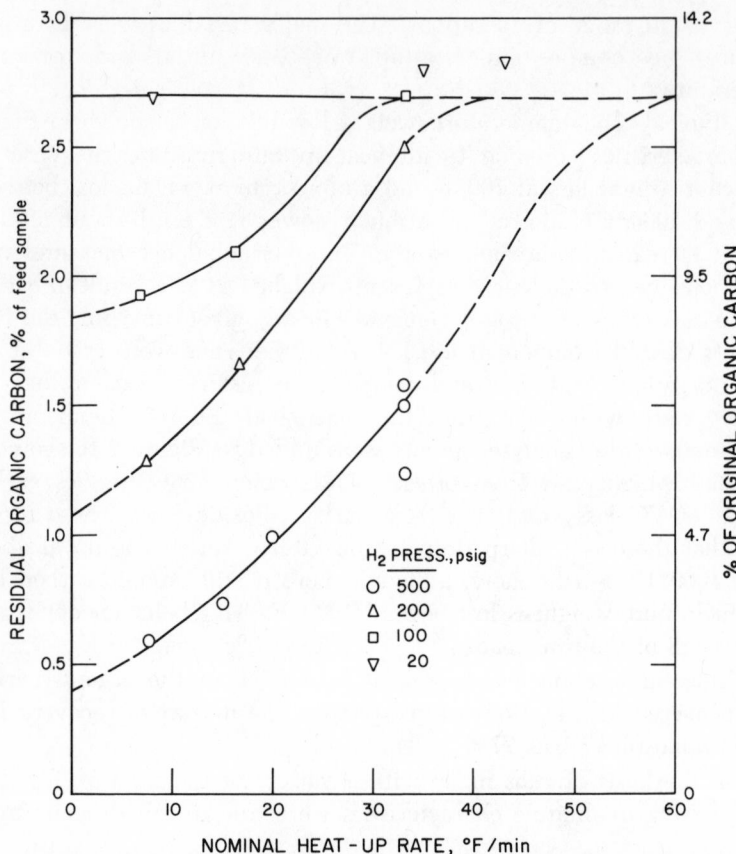

*Figure 3.    Pressure effect on kerogen recovery from rich shale in pure
hydrogen atmospheres*

tween residual organic carbon and residual carbon dioxide. As seen in
comparing the helium runs with the pure hydrogen runs, which have the
same time–temperature path to 1300°F, the former resulted in 23%
residual carbon and 80% residual carbon dioxide; the latter resulted in
7% residual carbon and 34% carbon dioxide. In two areas, this rough
correspondence does not hold. At low hydrogen pressure or very rapid
heat-up rates, the conditions are relatively poor for kerogen recovery
and very good for carbonate decomposition. Kerogen recovery also is
enhanced relative to carbon dioxide generation when the shale is soaked
in hydrogen at temperatures below 1000°F and above 700°F. Figure 4
shows that low carbonate decomposition (less than 20% of original)
can be achieved in simple heat-up paths but at the expense of leaving
8% of the shale's organic carbon. To achieve smaller amounts of residual

carbon, the most favorable path for minimal carbon dioxide generation is long soaking at 800°F before heating to higher temperatures. In that case, a residual carbon of only 3% of the original carbon can be achieved with 65% carbonate decomposition compared with almost 100% carbonate decomposition to achieve the same residual carbon by uniform heat-up.

An alternative method of suppressing carbonate decomposition is possible because the decomposition pressure of calcium carbonate is low enough to be overcome by a trace of carbon dioxide in the gas phase. When 2% carbon dioxide was added to the feed gas (10-psia carbon dioxide), the carbonate decomposition was much less than in the corresponding pure hydrogen runs.

However, while the carbonate decomposition was repressed, the kerogen recovery was also reduced relative to the pure hydrogen runs. Thus, the net improvement was not as striking as implied by the carbon dioxide repression (Figure 4).

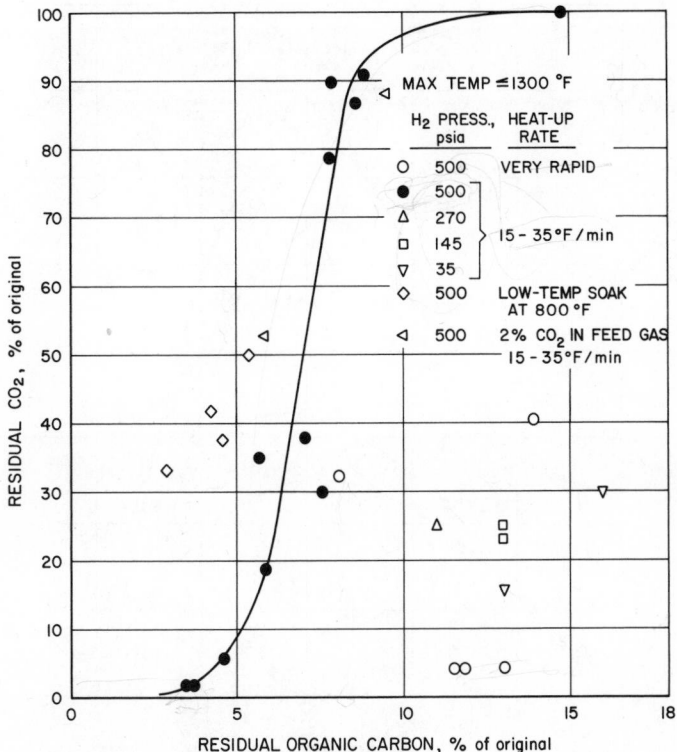

*Figure 4.    Correlation of rich shale kerogen recovery and carbonate decomposition in pure hydrogen*

**Results of Rich-Shale Studies—Kinetics.** LOW TEMPERATURE KERO-GEN RECOVERY. In the heat-up runs, there appears to be a significant decrease in weight-loss rate when the sample reaches about 1000°F and has lost about 25% of its original weight (Figure 2). In constant temperature runs below 1000°F, there appears to be a limiting weight loss, again about 25% of the rich shale. In this discussion, this portion of the recovered kerogen (and any other materials that comprise this weight loss) is referred to as the low temperature weight loss or kerogen recovery.

In Figure 5, the low temperature weight loss rates for several heat-up runs are shown as functions of the extent of weight loss. The range of operating conditions includes 15°–35°F/min temperature rise rates and 20–500 psig pressures. In all cases, the low temperature rate can be

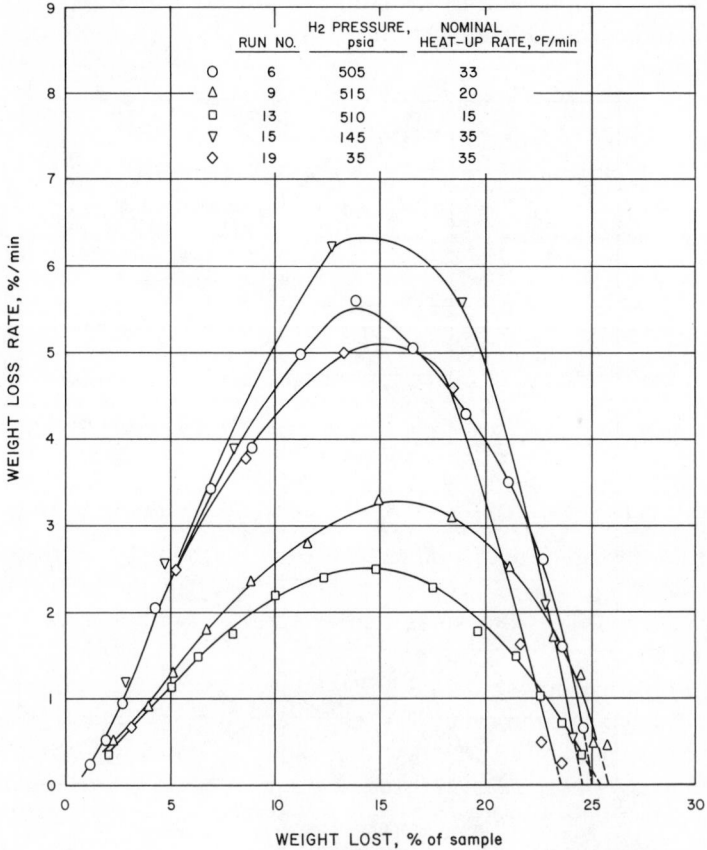

*Figure 5. Low temperature weight loss of rich shale (heat-up runs)*

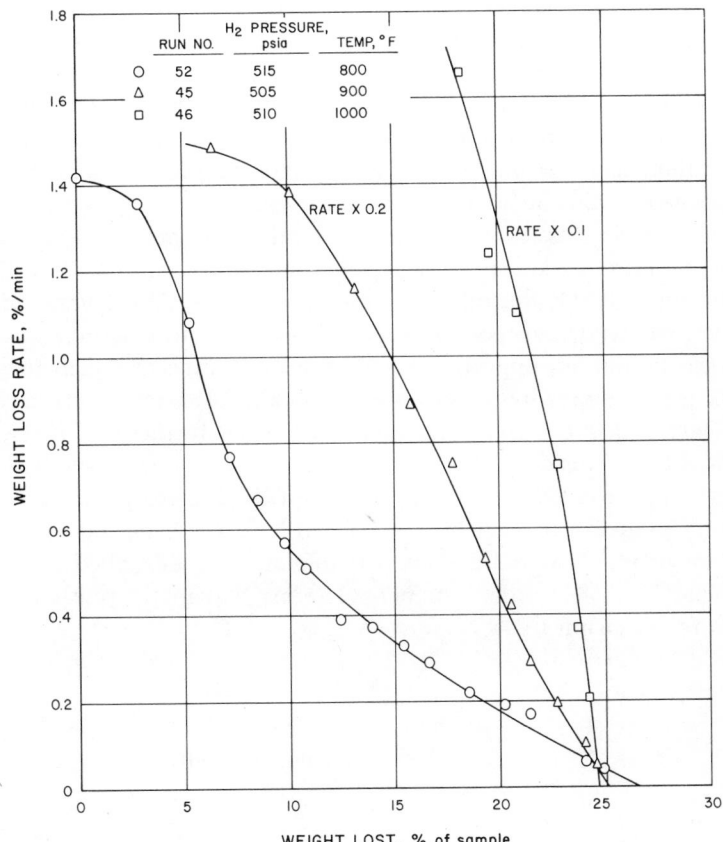

*Figure 6.   Low temperature weight loss of rich shale (constant-temperature runs)*

extrapolated to zero in the region of 23–25% weight loss. The actual rate goes through a minimum in this region (which also corresponds to 1000°–1100°F), the subsequent rise resulting from the carbon dioxide generation that begins at these temperatures and increases very rapidly (Figure 2).

In Figure 6, similar data are presented at higher conversions for the constant temperature runs at 800°, 900°, and 1000°F. Again, the rate of weight loss appears to become zero at about 25% weight loss. However, at 700°F the extrapolated weight lost at a zero loss rate was significantly lower than in the other runs, being about 11% rather than 25%.

It is apparent that, at temperatures of at least 800°F and, at most, 1000°F, a definite fraction of the kerogen can be removed and will be removed if enough time in a hydrogen atmosphere is allowed. Furthermore, if any more kerogen were to be recovered at temperatures below 1000°F, it would require additional time that is an order of magnitude greater than that needed for the low temperature kerogen recovery.

In the case of the rich shale, the low temperature weight loss, on the average, is 25.0% and (as are all kinetic data) is based on thermobalance data rather than on chemical analysis. There was no way to identify the components in this 25% from the data taken.

The randomness of the extrapolated values of about 25%, regardless of temperature, makes it safe to assume that no portion of that 25% is carbon dioxide. Of the other volatile components, 25% of the 28% in the original feed may constitute the low temperature kerogen, leaving 3%. This value is supported by the results of runs in which the shale was at temperatures high enough to achieve full low temperature weight loss (900°–1000°F), but inadequate to go any further. In these runs, the final OVC was 3.3–3.4%.

Attempts to devise a kinetic scheme for the low temperature kerogen recovery in hydrogen have been made but have not been successful. The constant temperature data appear capable of being described by a linear combination of first-order exponential terms that might result from the following mechanism:

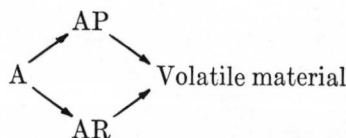

where AP and AR represent two different intermediate species. The determination of the appropriate kinetic parameters from the data proved to be very sensitive, and a satisfactory fit of the data has not been achieved.

In all aspects, the kinetics and total volatilization at high pressures in hydrogen below 1000°F are significantly different from those at pyrolysis conditions, exemplified by the work of Allred (3). In pyrolysis, the total volatilization below 1000°F is at least 10% less than that in high pressure hydrogen. A temperature of at least 880°F is needed to achieve complete volatilization under pyrolytic conditions, probably 100°F more than that in high pressure hydrogen. At constant temperature, the rate of volatilization in pyrolysis reaches a maximum at 50% of possible conversion according to Allred (3). In high pressure hydrogen, if there is a maximum, it occurs within the heat-up period ($\sim$ 1 min) and at less than 20% conversion. In high pressure hydrogen, the initial devolatilization rates are at least twice the maximum rates in pyrolysis reported by Hubbard and Robinson (4). These differences indicate strongly a significant difference in mechanism as well.

HIGH TEMPERATURE KEROGEN RECOVERY. The thermobalance data permit examination of the kinetics at low temperatures (1000°F) because of the fortuitous fact that no carbon dioxide is generated in that region. Above 1000°F, the carbon dioxide generation precludes any kerogen-generation rate data. Any analysis of the high temperature kerogen recovery must be based solely on the residue analyses.

Among the runs with rich shale in hydrogen, a surprisingly large number resulted in residual organic carbon of about 2.7% of the original sample. These included: instantaneous heat-up runs that did not have a reasonably long time below 1000°F, runs at low hydrogen pressures, and constant temperature runs below 1000°F. Runs in which temperatures higher than 1000°F were reached in hydrogen atmospheres left this amount or less residual carbon. It is reasonable to identify the 2.7% residual carbon of the rich shale with the residual of the low temperature kerogen recovery.

Eighteen runs at 500 psig with the rich shales resulted in residual carbon contents definitely lower than 2.7%. Simple relationships between the residual carbon and temperature history were sought but not found. One characteristic of the results is that comparable conversions can be obtained in shorter time periods if the sample is uniformly heated from 600° to 1300°F rather than being rapidly heated to 800°F or higher, held there, and then heated to 1300°F. This, together with the fact that we could not obtain high conversion (of the potential 2.7% high temperature product) with long times at high temperatures, implies that competing reactions are involved—one to the volatile product or its precursor and the other to a nonvolatilizable carbon. To even qualitatively explain both heat-up and constant temperature run results, one must assume that the basic competing reactions are fast and that the ultimate conversion is determined within a few minutes of exposing the shale to reaction temperatures (probably above 600°F).

**Results of Lean-Shale Studies.** Because of the small fraction of organic carbon in the original sample and the correspondingly small amount of carbon in the residue, it is difficult to be sure of the significance of the differences in the amounts of residual carbon among the products of the runs. Certainly, qualitatively, there are many similarities to the results with the rich shale such as:

1. About twice as much residual carbon is obtained in an inert gas (helium) as in hydrogen.

2. Without low temperature exposure, 89% of the organic carbon can be recovered compared with 87% for the rich shale.

3. Seven percent of the carbon remains in fast heat-up runs and 3% in slow heat-up runs. The rich shale showed 7.3% and 3.6%, respectively.

4. Carbon dioxide generation is suppressed by carbon dioxide in the feed gas, but with a penalty in kerogen recovery.

There is other behavior that is different. Most noticeable is a much lower carbon dioxide generation rate in the lean shale. Only very long exposures at high temperatures resulted in calcium carbonate decomposition. In most of the runs only 65% of the magnesium carbonate decom-

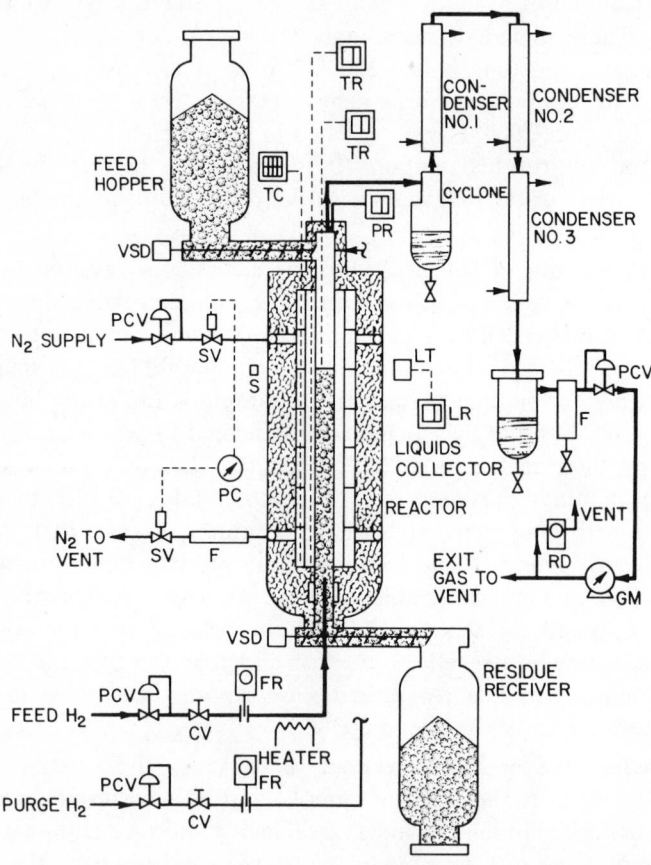

*Figure 7.   Flow and instrumentation diagram for high temperature, balanced pressure bench-scale unit. LR, level recorder; LT, level transmitter; TR, temperature recorder; PR, pressure recorder; FR, flow recording; RD, densitometer; PC, pressure controller; TC, temperature controller; GM, gas meter; F, filter; VSD, variable speed drive; PCV, pressure regulator; CV, control valve; SV, solenoid valve; S, gamma ray source.*

posed while in the rich shale all of it decomposes if the sample reaches 1300°F.

In the fast heat-up runs with lean shale, a smaller fraction of kerogen is recovered above 1000°F than one would expect from the rich-shale work, implying a difference in the low-temperature kerogen kinetics.

### Bench-Scale Tests

A flow diagram of the bench-scale unit is shown in Figure 7. The unit consists primarily of a hydrogasification reactor with associated equipment for feeding oil shale and hydrogen, for measuring their flow rates, and for collecting and measuring the quantities of residual shale, liquid products, and product gas. Simple controls were used to maintain reactor temperatures at the desired values, to maintain feed rates at constant values, to maintain constant reactor pressures, and to collect representative samples of the feed and product gases.

The reactor (Figure 8) consists of a cold-pressure shell, with a 24-in. id, and is about 23 ft long, containing an Incoloy 800 internal reactor tube (4-in. ips, schedule 40) which is heated by a seven-zone electric heater. The details of the design and operation of this reactor have been described previously (5). Each individually controlled zone is 31 in. long and has an id of about 7 in. The reactor pressure and the bed pressure drop are recorded continuously as are the reactor temperatures. A total of 32 thermocouples are attached to the outer tube wall, and several internal thermocouples were installed during the test program. These temperatures were recorded continuously along with the temperatures of the other pieces of equipment.

Shale was fed by a screw feeder from a pressurized hopper, which was filled before testing with sufficient shale for the entire run. The cold shale entered the top of the reactor and was preheated by countercurrent contact with the hot product gas. The residue shale was discharged from the bottom of the reactor into a pressurized residue receiver by a second screw feeder. The residue shale was cooled by countercurrent contact with cold feed hydrogen. In a few later runs at high hydrogen rates, the feed hydrogen was preheated.

The feed hydrogen entered the bottom of the reactor at a point above the discharge screw through a dip tube (Figure 8). The exit gas left through a dip tube extending down into the top of the reactor. In later runs, a sintered metal bayonet filter was installed on the end of the dip tube to remove dust and eliminate occasional plugging of the outlet lines. The feed shale dropped down through a third dip tube that served to keep any heavy liquids in the exit gas from condensing on the feed shale and plugging the tube at the feed screw outlet.

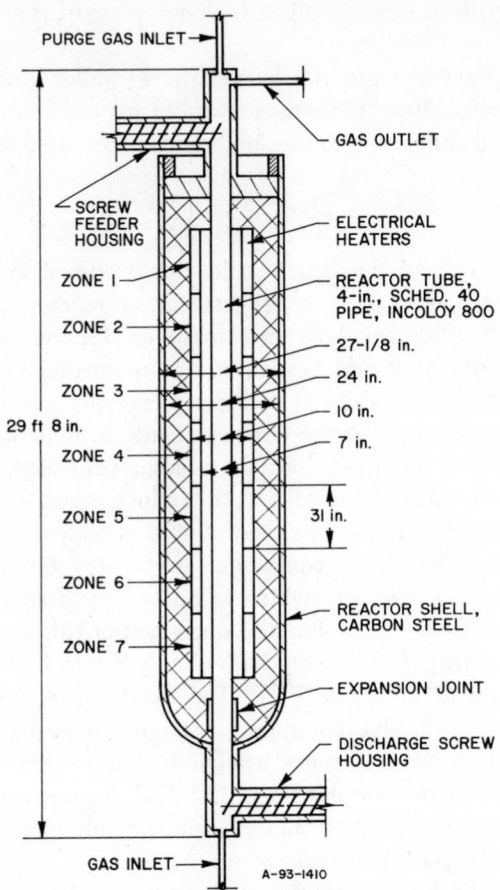

PURGE GAS INLET→

GAS OUTLET

SCREW
FEEDER
HOUSING

ELECTRICAL
HEATERS

ZONE 1

REACTOR TUBE,
4-in., SCHED. 40
PIPE, INCOLOY 800

ZONE 2

27-1/8 in.

24 in.

ZONE 3

10 in.

29 ft 8 in.

7 in.

ZONE 4

ZONE 5

31 in.

ZONE 6

REACTOR SHELL,
CARBON STEEL

ZONE 7

EXPANSION JOINT

DISCHARGE SCREW
HOUSING

GAS INLET→

A-93-1410

*Figure 8.   High temperature, balanced pressure
reactor*

Hydrogen feed gas and hydrogen purge gas rates were both measured by orifice meters. The shale was fed constantly at the desired rate,
and the rate of shale discharge was manually adjusted to give a constant
shale bed level. The shale bed level was indicated by a nuclear type
level indicator and was recorded continuously. More accurate measures
of the shale feed and discharge rates were obtained by weighing the
shale initially charged to the feed hopper and that present in the feed
hopper and residue receiver at the end of the run. Feed and residue
shales were analyzed by methods described in Appendix B of Ref. *1*.
The exit gas volume was measured by a conventional iron case type
meter. Aliquot samples of the exit gas were collected during the steady-
state part of each run. Gas samples were analyzed by mass spectrometer

and gas chromatograph. The exit gas specific gravity was monitored continuously by a recording gravitometer. The product liquids were collected in two different knockout pots and analyzed by conventional ASTM procedures. The first knockout pot (operated at high temperature) contained a short cyclone section which removed dust and heavy tar. The exit gas was then cooled in a three-stage condenser to condense out low temperature liquids which were collected in a low temperature knockout pot. Both pots were drained, and the collected liquids were weighed every 1/2 hr during the steady-state portion of each run.

All of the tests conducted in the pilot plant were with a single batch of Colorado oil shale from the U.S. Bureau of Mines mine at Rifle, Colo. Mine run material was crushed to 1/4-in. size at the mine and sieved elsewhere into various fractions. We selected a −6+10 U.S. Standard sieve size material for most of these tests because it was the largest size which we could successfully feed with our existing equipment. This material had a Fischer assay oil yield in the 20–25 gal/ton range.

**Analysis of Oil Shale.** A typical analysis of the oil shale used in the tests reported here is given in Table II.

**Effects of Operating Temperature.** The first series of tests was conducted to study the effects of reaction zone temperature on the yields and properties of gaseous, liquid, and solid products. These tests were

**Table II.    Typical Analysis of Colorado Oil Shale
Used in Bench-Scale Test Program**

| | |
|---|---:|
| Moisture (wt %) | 0.30 |
| Composition (wt %, dry basis) | |
|     organic carbon | 11.30 |
|     carbon dioxide | 17.55 |
|     hydrogen | 1.72 |
|     nitrogen | 0.35 |
|     oxygen (by difference) | 1.73 |
|     sulfur | 0.54 |
|     ash | 66.81 |
|     total | 100.00 |
| Screen analysis, U.S. standard sieve size (wt %) | |
|     +6 | 0.1 |
|     +8 | 12.5 |
|     +16 | 79.9 |
|     +30 | 6.9 |
|     +60 | 0.3 |
|     +100 | 0.1 |
|     +200 | 0.1 |
|     −200 | 0.1 |
|     total | 100.0 |

conducted with relatively short (9–12 ft) beds and flat temperature profiles. The results of these tests are summarized graphically in Figures 9 and 10.

As expected, gaseous hydrocarbon yields and mineral carbonate decomposition increase with increases in temperature (Figure 9). The results are very encouraging—in all tests, over 90% of the organic carbon was recovered as gaseous and liquid products. At 1200°F, over 80% of the organic carbon was converted to liquids; at 1400°F, over 60% was converted to gas.

The effect of temperature on kerogen conversion is not pronounced although some increase can be noted in the neighborhood of 1300°F. The test at 1200°F was conducted with a shale space velocity that was considerably higher than that used in the tests at 1300° and 1400°F.

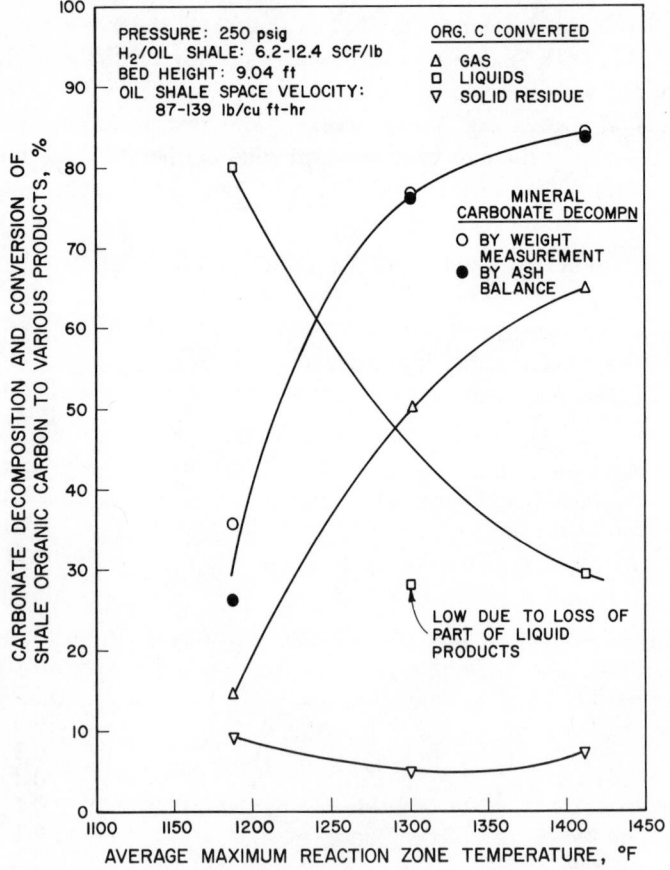

*Figure 9.   Product distribution in bench-scale tests*

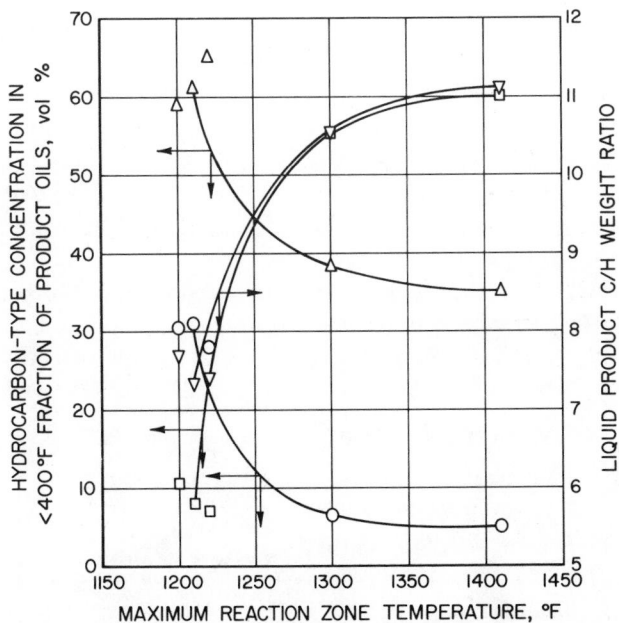

*Figure 10. Effects of reaction zone temperature on liquid product properties. Hydrocarbon type—○, saturates; △, olefins; □, aromatics. C/H weight ratio—▽.*

More organic carbon could probably have been removed at 1200°F if a longer residence time had been provided. This higher space velocity probably also caused a relatively lower mineral carbonate decomposition.

The effects of temperature on the lighter ( < 400°F) liquid product properties are shown in Figure 10. Although less hydrocarbon liquids are formed at the higher temperatures, those formed would be more difficult to gasify. First, the saturates plus olefins content of the IBP (initial boiling point) to 400°F fraction decreases from about 92 vol % to about 40 vol % as the temperature increases from 1200° to 1400°F. Also the carbon/hydrogen weight ratio of the entire liquid product increases from an average of about 7.5 to about 11 as the temperature increases from 1200° to 1400°F. Thus, at the higher temperatures, an increasing portion of the liquid product is aromatic. Of course the lighter aromatics, such as benzene, toluene, and xylene, are very difficult to gasify. The heavier aromatics, however, will produce additional hydrocarbon gases with further reaction. The hydrocarbon type analysis could be performed only on the < 400°F fraction of the oils, and in no run did the < 400°F fraction constitute more than 35% of the total oil.

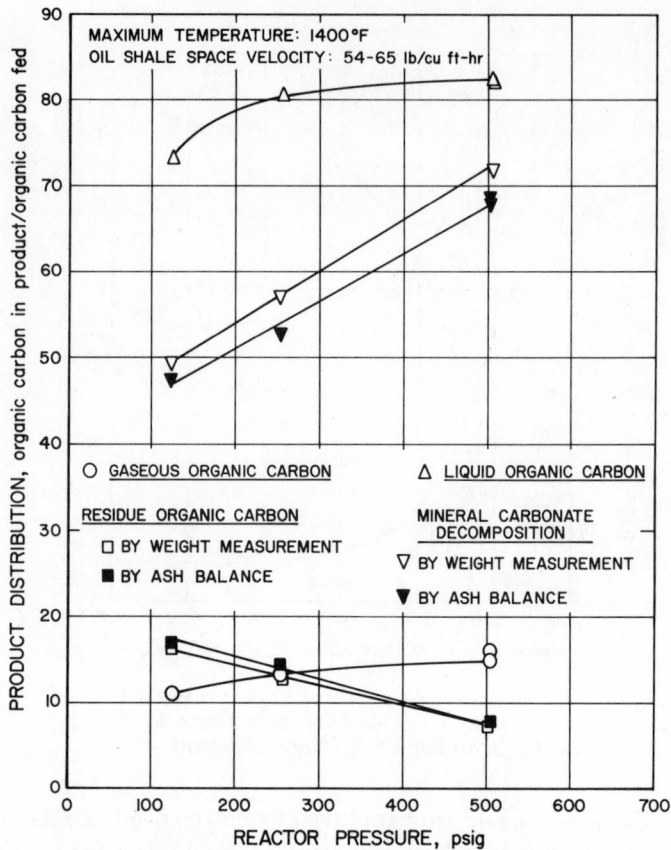

*Figure 11.  Effect of pressure on organic carbon distribution in products and mineral carbonate decomposition*

The liquid products produced in these tests were of substantially better quality than those produced by conventional retorting. In all of these tests, at least 89% boiled below 730°F, whereas in conventional retorting processes reported in the literature, less than 50% boiled below 800°F.

Because bed heights of about 9–12 ft (3½–5 heating zones) were used, the feed shale fell through a long heated zone (of 1000°F or more) before reaching the shale bed. This gave an initial rapid heat-up of the shale feed rather than the slow heat-up which have been preferred based on laboratory thermobalance tests. However, based on heat transfer calculations, we estimate that the shale particles reached an average temperature of no greater than about 900°–1000°F by the time they reached the shale bed. Further heating to final bed temperature ranged from about 20° to 30°F/min, which is favorable to higher yields.

**Effects of Operating Pressure.** A second series of tests was conducted to study the effects of reaction zone pressure on the yields and properties of each product. These tests were conducted with a relatively steep temperature gradient in the bed (to prehydrogenate the shale) and bed depths ranging from 12.9 to 15.5 ft (5–6 zones). In Figure 11, the product yields are shown as functions of pressure. The product gas yield is only slightly affected by pressure. The liquid products, however, increase with pressure. Thus, the organic carbon remaining in the residue shale decreases with pressure. More than twice as much remains at 125 psig. The reason for the lower gas yields and higher amounts of residual organic carbon in these runs (compared with previous runs at a 1400°F maximum temperature) is a shorter residence time at high temperature. In these tests, we used a steeper temperature gradient in the upper part of the bed to prehydrogenate the feed shale.

The effects of pressure on hydrocarbon types in the liquid products (400°F fraction) and the carbon/hydrogen ratio in the total oil are

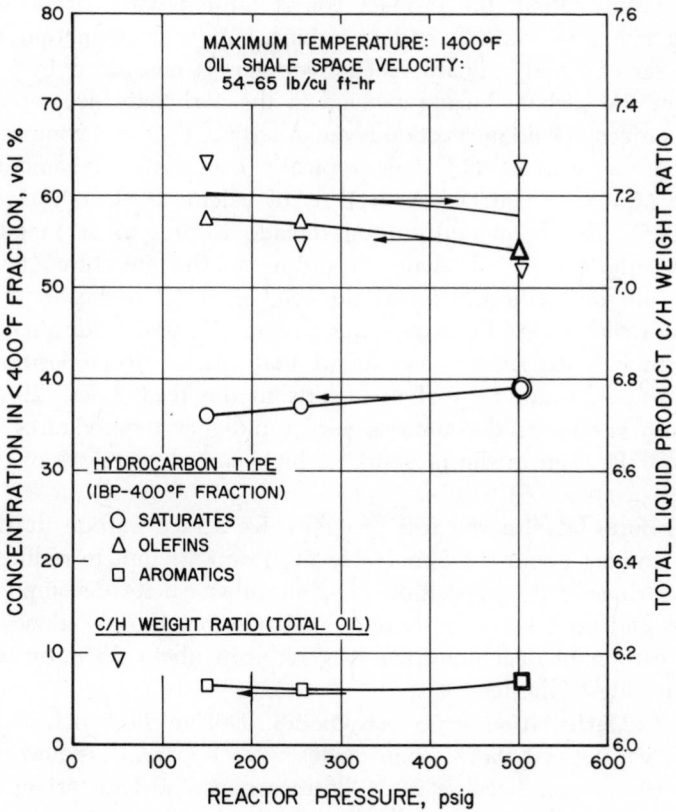

*Figure 12. Effect of pressure on liquid product properties*

shown in Figure 12. There are only slight effects in the range studied. The olefins fraction decreases and the saturates fraction increases with increases in pressure, but the effect is slight. There is no discernible effect of pressure on the aromatic fraction. The carbon/hydrogen ratio of the total liquid products also is not significantly affected by pressure. This agrees with the fact that the olefins fraction also remains nearly constant. One would expect a lower carbon/hydrogen ratio at higher pressures because of hydrogenation of the oils. The absence of any oil hydrogenation is probably caused by the low temperatures in the empty space above the bed and within the upper part of the bed, so that the primary liquid products are not hydrogenated.

**Mineral Carbonate Decomposition.** In tests to show the effects of temperature on the hydrogasification behavior of oil shale, mineral carbonate decomposition ranged from about 26% at a 1200°F maximum reactor temperature to about 85% at a 1400°F maximum reactor temperature. High mineral carbonate decompositions are undesirable for several reasons. First, the product gas is diluted with carbon oxides, requiring extensive shifting and scrubbing and/or methanation of the product gas. Second, valuable feed hydrogen is consumed by reverse shifting of the carbon dioxide evolved in the carbonate decomposition. Third, the decomposition reaction is endothermic, thus removing sensible heat from the system. Oil shale typically contains large amounts of dolomite ($CaCO_3 \cdot MgCO_3$) as well as of calcite ($CaCO_3$). Approximately 36% of the mineral carbon dioxide is present as magnesium carbonate in Colorado oil shales according to the literature (6). We analyzed our oil shale and found the feed material to be about 47% magnesium carbonate. From previous studies (1) and from our laboratory thermobalance studies, we found that calcite decomposition can be suppressed by adding carbon dioxide to the feed gases. However, magnesium carbonate decomposes very rapidly at temperatures above about 1000°F, even in the presence of high partial pressures of carbon dioxide.

Therefore, later in the test program, we added carbon dioxide to the feed gas (at about the 5-mol% level) in an attempt to reduce total mineral carbonate decomposition (*i.e.*, suppress calcite decomposition) since this had been successfully done in laboratory thermobalance tests. Mineral carbonate decomposition was reduced about 25% by adding carbon dioxide to the feed gas.

***In Situ* Methanation of Carbon Oxides.** During the bench-scale test program, we noticed that carbon oxides were apparently being methanated in the reactor because more hydrocarbons and less carbon oxides were present in the products than would be indicated by carbon and carbon oxides balances. We then carried out two tests with no oil shale present in the reactor and with feed gases containing from about 4–6 mol% carbon dioxide at temperatures and pressures in the region used in the oil shale hydrogasification tests. In the first test, we simply passed the feed gas through the reactor tube packed with sand. We found that about 59% of the carbon dioxide was converted to gaseous hydrocarbon (methane and ethane). Because it was possible that the metal reactor tube was catalyzing the reaction rather than the solids, we conducted a second test with an empty reactor tube to ensure that the methanation observed in our bench-scale tests indicated what would happen in a large scale plant, where the reactor would be refractory-lined and reactor gases would not contact a metallic wall. In this test, only about 18% of the carbon dioxide was converted to gaseous hydrocarbons, indicating that the metal tube wall was only partially contributing to the observed methanation. Since the gas residence time and gas contact with the metal wall were greater in the latter test than in the first test, apparently even less of the observed methanation in the bench-scale tests is the result of catalysis by the wall.

We also made thermodynamic equilibrium calculations to see how much methane could be formed by methanation of carbon oxides. These calculations show that, for all conditions within the reactor tube, it is possible (thermodynamically) to form methane in quantities even greater than those observed.

### Summary and Conclusions

Results of the small-scale laboratory thermobalance studies have shown that:

1. The presence of hydrogen, even at low pressure, significantly increases organic carbon recovery compared with hydrogen-free retorting.

2. Good organic carbon recovery can be achieved at an elevated hydrogen partial pressure. At 500 psia, slow heating can achieve over 95% recovery.

3. The heating rate significantly affects organic carbon recovery. Very rapid heating to 1300°F limits recovery to about 87%.

4. The carbon dioxide generated by decomposition of calcium carbonate and magnesium carbonate begins at about 1000°F, increasing rapidly above this temperature. Decomposition of the calcium carbonate can be almost completely suppressed by adding small amounts of carbon dioxide in the feed gas stream; magnesium carbonate decomposition cannot be suppressed at hydrogasification temperatures.

Results obtained in the 4-in. diameter reactor have generally verified the trends observed in the laboratory study. Kerogen recoveries using hydrogen have exceeded 90% which is significantly better than recoveries obtained in conventional retorting.

Constant temperature runs (rapid shale heat-up in free fall above the bed) indicate that a high temperature favors the production of gas and promotes aromatization of the liquid products. The use of controlled shale heat-up (shale preheated in the upper portion of the bed) favors liquid production; however, the liquid products are of much higher quality.

Additional data obtained in the countercurrent tests verified the ability to suppress carbonate decomposition by adding carbon dioxide to the feed hydrogen. We also discovered that substantial methanation of carbon oxides occurs in the shale bed.

In summary, the experimental program has shown that the basic concept of countercurrent operation is technically feasible. The overall chemistry of the system is now better defined and looks very favorable.

### Acknowledgment

The work described was sponsored by the American Gas Association. The authors wish to express their appreciation to the Association for permission to publish this paper. We also wish to thank H. W. Sohns and M. D. Smith of the U.S. Bureau of Mines who provided us with the oil shale used in this program.

### Literature Cited

1. Feldman, H. F., Bair, W. G., Feldkirchner, H. L., Tsaros, C. L., Shultz, E. B., Jr., Huebler, J., Linden, H. R., "Production of Pipeline Gas by Hydrogasification of Oil Shale," *Inst. Gas Technol. Chicago, Res. Bull.* (1966) **36.**
2. Feldkirchner, H. L., Johnson, J. L., "High-Pressure Thermobalance," *Rev. Sci. Instrum.* (1968) **39,** 1227.
3. Allred, V. D., "Kinetics of Oil Shale Pyrolysis," *Chem. Eng. Prog.* (1966) **62,** 55.
4. Hubbard, A. B., Robinson, W. E., "A Thermal Decomposition Study of Colorado Oil Shale," *U.S. Bur. Mines Rep. Invest.* (1950) **4244.**
5. Dirksen, H. A., Lee, B. S., "Balanced-Pressure Pilot Reactor," *Chem. Eng. Prog.* (1966) **62,** 98.
6. Stanfield, K. E. *et al.,* "Properties of Colorado Oil Shale," *U.S. Bur. Mines Rep. Invest.* (1971) **4825.**

RECEIVED December 17, 1974.

# Production of Synthetic Crude from Crude Shale Oil Produced by *in Situ* Combustion Retorting

C. M. FROST, R. E. POULSON, and H. B. JENSEN

U.S. Energy Research and Development Administration, Laramie Energy Research Center, Laramie, Wyo. 82071

*Of six modern hydrogenation catalysts tested on a shale oil heavy gas oil, a Ni–Mo-on-alumina catalyst was found superior at 1500 psig and 800°F. This catalyst was used to prepare a premium synthetic crude oil from in situ combustion-produced crude shale oil by hydrogenation of naphtha, light oil, and heavy oil fractions. These fractions were obtained by distillation of the oil, coking the vacuum residuum, and combining like distillate ranges. The synthetic crude contained 250 ppm nitrogen and 5 ppm sulfur. Physical properties were: API gravity, 43.9°; pour point, 32°F; viscosity at 100°F, 32 SUS. The overall yield was 103 vol. %.*

Among the considerations associated with the development of a shale oil industry are the costs and environmental hazards of disposing of the spent shale that is inherently produced if there is any mining and aboveground retorting. For several years the Bureau of Mines has been engaged in research to recover shale oil by *in situ* combustion retorting, an approach that avoids spent shale disposal. Both actual underground (1–3) and simulated *in situ* retorting (4–7) are being investigated.

Crude shale oils produced by *in situ* combustion retorting of Green River oil shale normally have higher API gravities and lower viscosities and pour points than crude shale oils produced in N-T-U or gas combustion retorts (8). *In situ* crude shale oils also contain a much higher percentage of material boiling below 1000°F. While the nitrogen contents of *in situ* crude shale oils may be somewhat lower than those of crude shale oils produced in N-T-U or gas combustion retorts, they still contain more than twice as much nitrogen as high nitrogen petroleum crudes.

Since existing refineries would not be able to cope with the high nitrogen content of raw shale oil if it were a substantial part of the refinery feed, the National Petroleum Council (NPC) (9) suggested that crude shale oil be upgraded at the retorting site by a process of catalytic hydrogenation to produce a premium feedstock called "syncrude." In this process, the crude shale oil would be distilled to produce naphtha, light oil, heavy oil, and residuum. The residuum would be processed in a delayed-coking unit to produce petroleum coke and a vapor stream containing gas, naphtha, light oil, and heavy oil. Vapor from the coking unit would flow back to the crude distillation facilities for separation into various fractions. The naphtha, light oil, and heavy oil would be subsequently hydrogenated to remove nitrogen and sulfur and to reduce the viscosity and pour point of the finished syncrude.

There is some disagreement in the literature as to which type of hydrotreating catalyst most effectively removes nitrogen from crude shale oil and/or shale oil coker distillates. Carpenter and Cottingham (10) found that, of 17 catalysts tested, a cobalt–molybdate-on-alumina catalyst was superior for removing nitrogen from crude shale oil. Benson and Berg (11) reported that, of 12 catalysts tested, an HF-activated cobalt–molybdate catalyst was superior for removing nitrogen from shale oil coker distillates. On the other hand, Montgomery (12) reported that catalysts containing high concentrations of nickel and tungsten were best for hydrodenitrogenation. These investigators did not report any work with nickel–molybdenum or nickel–cobalt–molybdenum catalysts.

The purpose of the present study was to test various modern hydrogenation catalysts for their effectiveness in removing nitrogen from in situ crude shale oil fractions and to determine the feasibility of producing a synthetic crude oil that would meet the specifications for syncrude suggested by the NPC. Six modern hydrogenation catalysts were tested to compare their efficiencies in removing nitrogen from a heavy (600°1000°F) in situ gas oil. The best catalyst, nickel–molybdenum-on-alumina, was used in preparing fractions of a synthetic crude using the methods suggested by NPC.

### Properties of in Situ Crude Shale Oil

The crude shale oil used in this study was obtained from an in situ combustion retorting experiment at Rock Springs, Wyo. (1, 2), during the last week of the experiment and is considered a representative "steady state" oil. Properties of the in situ crude shale oil are shown in Table I.

### Table I. Properties of *in Situ* Crude Shale Oil

| | |
|---|---|
| Gravity (°API) | 28.4 |
| Nitrogen (wt %) | 1.41 |
| Sulfur (wt %) | .72 |
| Pour point (°F) | 40 |
| Viscosity (SUS at 100°F) | 78 |
| Carbon residue (wt %) | 1.7 |
| Ash (wt %) | .06 |

## Experimental Procedures and Results

**Apparatus and Operating Procedure.** A simplified flow diagram of the hydrogenation unit is shown in Figure 1. The reactor was a 40-in. long type 316 stainless steel tube. The catalyst bed was supported by a stainless steel screen 11 in. from the bottom of the reactor. A second screen was placed at the top of the catalyst bed, and the upper part of the reactor was filled with quartz chips and served as a preheater for oil and hydrogen. The reactor was surrounded by an aluminum block with a 3-in. od and a 1-in. id and was heated by a four-zone electric furnace, each zone of which was independently controlled. Temperatures were measured by five thermocouples placed in a groove in the aluminum block adjacent to the reactor and spaced at equal intervals along the length of the catalyst bed and preheater. With proper adjustment of the heating elements, the recorded temperatures could be maintained within 5°F of each other.

At the beginning of each experiment the catalyst was heated to 700°F with air passing through the reactor at the rate of 1.1 standard cu. ft per pound of catalyst per hr (scf/lb/hr). Steam was then introduced at the rate of 0.41 pound per pound of catalyst per hr (lb/lb/hr), and these conditions were maintained for 16 hr. The reactor was then

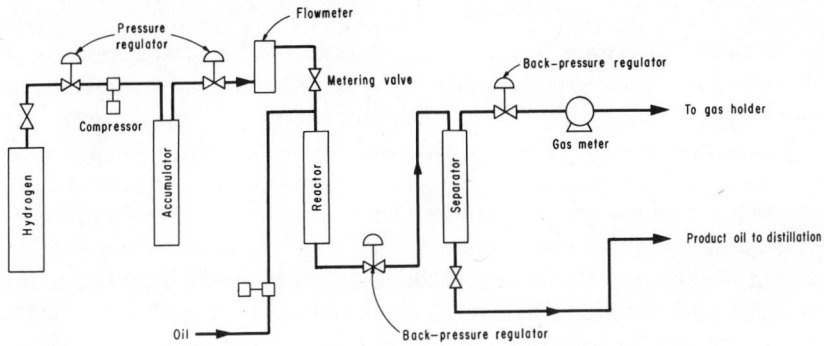

*Figure 1. Simplified flow diagram of hydrogenation unit*

cooled to 500°F, and the steam was cut off. After the system cooled to 350°F, the air flow was stopped. The reactor was then purged with helium and pressurized to 250 psig. A hydrogen stream containing 5 wt % hydrogen sulfide was passed through the reactor at a rate of 1.0 scf/lb/hr for 3 hr to sulfide the catalyst. The temperature, pressure, and hydrogen flow rate were adjusted to those required for the particular experiment, and the oil flow was started. Hydrogen (99.9% purity) was used directly from standard shipping cylinders without further purification.

Products from the reactor passed through a back-pressure regulator into a separator maintained at 75°F and 200 psig. Tail gas from the separator passed through a second back-pressure regulator and was metered and sampled. Liquid products were drained from the separator after each 24-hr period of operation and washed with water to remove ammonia and hydrogen sulfide before a sample was taken for analysis. At the conclusion of each experiment the oil and hydrogen flows were stopped, the reactor was depressurized, and steam was introduced at the rate of 0.41 lb/lb/hr while the reactor was cooled to 700°F. Air was then introduced at the rate of 1.1 scf/lb/hr, and these conditions were maintained until the coke burnoff was completed.

## Table II.   Catalysts Tested

| Manufacturer | Manufacturer's No. | Active Metals | Catalyst Designation |
|---|---|---|---|
| American Cyanamid Co. | Aero HDS–2A | Co–Mo | I (Co–Mo) |
| American Cyanamid Co. | Aero HDS–3A | Ni–Mo | II (Ni–Mo) |
| Harshaw Chemical Co. | Ni 4301–E | Ni–W | III (Ni–W) |
| Harshaw Chemical Co. | Ni 4303–E | Ni–W | IV (Ni–W) |
| Nalco Chemical Co. | NM–502 | Ni–Mo | V (Ni–Mo) |
| Davidson Chemical Co. | NICOMO | Ni–Co–Mo | VI (Ni–Co–Mo) |

**Catalyst Screening Tests.** The catalysts used in these tests were obtained from commercial sources. Table II shows the manufacturer, manufacturer's number, active metals, and catalyst designation for each of the catalysts tested. Three of the catalysts were received in the form of 1/16-in. extrusions and were used as such. The other catalysts, obtained in larger sizes, were crushed and sized to 10–20 mesh.

Charge stock for the catalyst testing experiment was prepared by topping a sample of the *in situ* crude shale oil to 600°F in a batch still equipped with a column having 35 trays and then separating the 600°–1000°F fraction in a vacuum flash distillation unit. Properties of the 600°–1,000°F fraction of *in situ* crude shale oil are shown in Table III.

**Table III. Properties of 600°–1000°F Distillate from**
**in Situ Crude Shale Oil**

| | |
|---|---|
| Gravity (°API) | 23.3 |
| Nitrogen (wt %) | 1.66 |
| Sulfur (wt %) | .51 |
| Viscosity (SUS at 100°F) | 111 |
| Carbon residue (wt %) | .5 |
| Distillation | |
| initial boiling point (°F) | 579 |
| % recovered | |
| 5 | 618°F |
| 10 | 620°F |
| 20 | 648°F |
| 30 | 666°F |
| 40 | 681°F |
| 50 | 699°F |
| 60 | 732°F |
| 70 | 768°F |
| 80 | 809°F |
| 90 | 867°F |
| 95 | 914°F |
| end point | 995°F |

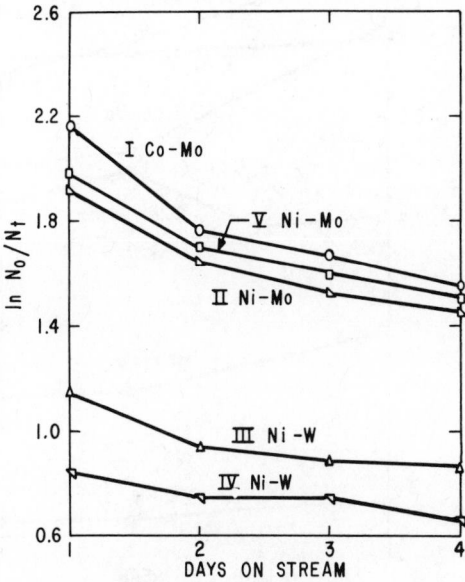

*Figure 2. Effect of operating time on denitrification rate constants at 1000 psig*

All catalyst tests were run at an operating temperature of 800°F, a space velocity of 1.0 wt of oil per wt of catalyst per hr ($W_o/W_c/hr$), and a hydrogen feed rate of 5000 standard cu ft per barrel of feed (scf/bbl). Eighteen grams of catalyst were charged to the reactor in each test. Five of the six catalysts were tested at an operating pressure of 1000 psig, and all six were tested at 1500 psig. Each test was continued for 96 hr.

In Figure 2 the first-order denitrification rate constants at 1000 psig expressed as the logarithm of the ratio of nitrogen in the feed to nitrogen in the liquid product ($\ln N_o/N_t$) are plotted against days on-stream to show the effect of operating time on catalyst activity. These curves show that the activities of all the catalysts tested decreased quite rapidly with time on-stream. The catalyst with the highest denitrification activity at an operating pressure of 1000 psig was catalyst I (Co–Mo), and the one with the lowest denitrification activity was catalyst IV (Ni–W).

Figure 3 shows a plot of the first-order denitrification rate constants *vs.* time on-stream at an operating pressure of 1500 psig. The catalysts with the highest initial denitrification activity at 1500 psig were catalyst

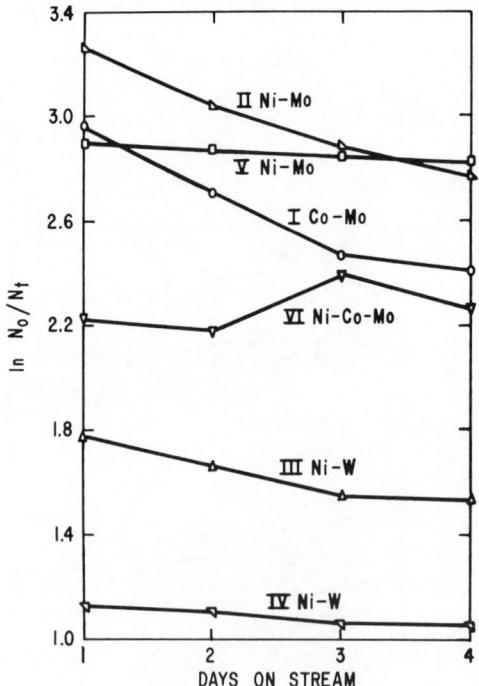

*Figure 3. Effect of operating time on de-nitrification rate constants at 1500 psig*

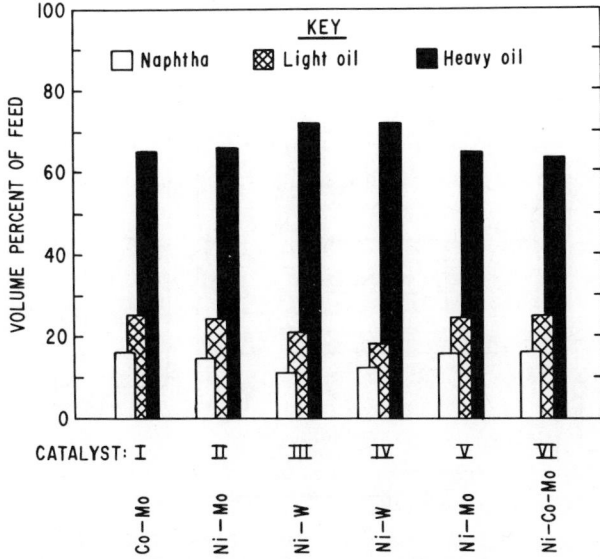

*Figure 4. Effect of catalyst on distillate yields at 1500 psig*

II (Ni–Mo) and catalyst I (Co–Mo). However, catalyst V (Ni–Mo) showed the highest activity after 96 hr, and its activity decreased very little with time on-stream. Catalyst VI (Ni–Co–Mo) also had a high resistance to deactivation but had a much lower initial activity. Catalyst IV (NiW) again showed the lowest denitrification activity.

Composite samples of the total liquid product from each test were fractionated to determine the degree of hydrocracking attained with each catalyst. Distribution of the various distillate fractions for the tests made at 1500 psig are shown in Figure 4. Yields of naphtha and light oil were uniformly lower for tests made at 1000 psig. The highest vol % yields of liquid product were attained with catalysts I (Co–Mo) and V (Ni–Mo) and the lowest yields with catalysts III (Ni–W) and IV (Ni–W). The highest conversion, *i.e.*, material converted to products boiling below 550°F, was attained with catalyst VI (Ni–Co–Mo). The lowest conversion was attained with catalyst IV (Ni–W), a hydrocracking catalyst. The highest yields of naphtha and light oil were attained with catalysts I (Co–Mo) and VI (Ni–Co–Mo). Because of its high sustained denitrification activity, catalyst V (Ni–Mo) was selected for use in the preparation of syncrude by hydrogenation of the *in situ* distillate fractions.

**Preparation of Synthetic Crude.** The overall flow diagram for upgrading crude shale oil is shown in Figure 5. A sample of *in situ* crude

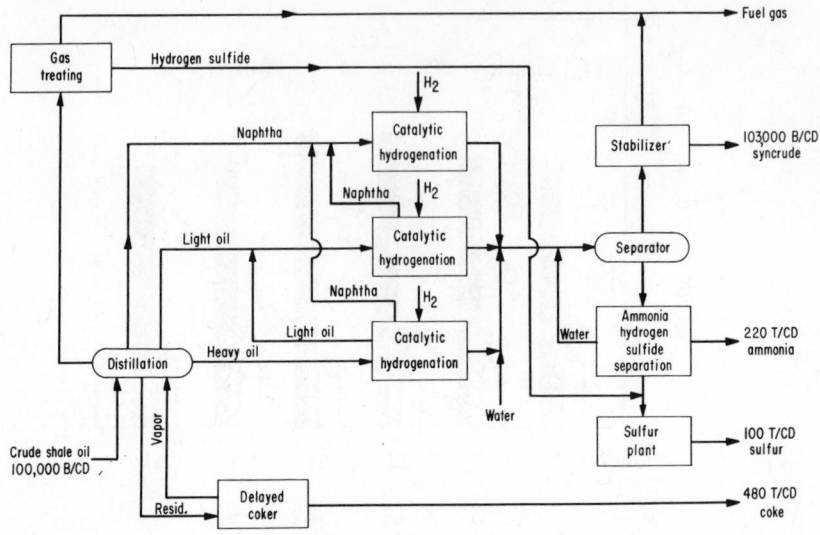

*Figure 5. Flow diagram for upgrading* in situ *crude shale oil*

shale oil was fractionated in a Podbielniak Hypercal distillation unit to obtain a 350°F end-point naphtha, a 350°–550°F light oil, a 550°–850°F heavy oil, and a vacuum residuum. Residuum from the fractionation was coked at atmospheric pressure in a laboratory coking unit, and the

### Table IV. Yields of Products from Distillation and Coking

| % of Crude | First Distillation | | Final Distillation | |
|---|---|---|---|---|
| | *Weight* | *Volume* | *Weight* | *Volume* |
| IBP–350°F | 6.20 | 6.86 | 6.30 | 7.00 |
| 350°–550°F | 43.71 | 45.50 | 37.58 | 39.24 |
| 550°–850°F | 28.89 | 28.60 | 34.62 | 33.50 |
| 850°F+ | 21.20 | 18.94 | 21.50 | 19.26 |
| Coker distillate | 17.21 | 17.50 | 17.30 | 17.47 |
| Coke | 2.83 | | 3.08 | |
| Hydrogen | .01 | | .02 | |
| Methane | .26 | | .27 | |
| Ethane | .17 | | .16 | |
| Ethylene | .02 | | .03 | |
| Propane | .22 | | .20 | |
| Propylene | .11 | | .09 | |
| Isobutane | .06 | .09 | .03 | .05 |
| Butane | .09 | .14 | .10 | .15 |
| Butenes | .09 | .13 | .10 | .14 |
| Carbon monoxide | .02 | | .03 | |
| Carbon dioxide | .03 | | .02 | |
| Hydrogen sulfide | .08 | | .07 | |

### Table V. Properties of Fractions from Distillation of *in Situ* Crude Shale Oil

|  | Original Crude | Final Distillation |
|---|---|---|
| C$_5$–350°F naphtha |  |  |
| gravity (°API) | 45.4 | 45.8 |
| nitrogen (wt %) | 1.16 | 1.09 |
| sulfur (wt %) | .52 | .70 |
| 350°–550°F light oil |  |  |
| gravity (°API) | 34.9 | 35.0 |
| nitrogen (wt %) | 1.24 | 1.29 |
| sulfur (wt %) | .52 | .54 |
| 550°–850°F heavy oil |  |  |
| gravity (°API) | 26.8 | 27.2 |
| nitrogen (wt %) | 1.53 | 1.61 |
| sulfur (wt %) | .46 | .46 |
| 850°F+ residue |  |  |
| gravity (°API) | 11.3 | 11.2 |
| nitrogen (wt %) | 1.93 | 1.91 |
| sulfur (wt %) | .60 | .56 |

### Table VI. Hydrogenation of 550°–850°F Heavy Oil

Operating conditions

| temperature (°F) | 800 |
|---|---|
| pressure (psig) | 1500 |
| space velocity (W$_o$/W$_c$/hr) | 1.0 |
| hydrogen feed (scf/bbl) | 5000 |
| hydrogen consumed (scf/bbl) | 1320 |

| Yields of Products | % of Feed | | % of Crude | |
|---|---|---|---|---|
|  | Weight | Volume | Weight | Volume |
| C$_5$+ liquid product | 95.20 | 101.68 | 39.55 | 41.94 |
| C$_5$–175°F | 1.92 | 2.52 | .80 | 1.04 |
| 175°–350°F | 9.25 | 10.77 | 3.84 | 4.44 |
| 350°–550°F | 23.75 | 24.92 | 9.87 | 10.28 |
| 550°F+ | 60.28 | 63.47 | 25.04 | 26.18 |
| coke | .12 |  | .05 |  |
| hydrogen | −2.24 |  | −.93 |  |
| methane | .77 |  | .32 |  |
| ethane | .87 |  | .36 |  |
| propane | .88 |  | .37 |  |
| isobutane | .30 | .47 | .12 | .19 |
| butane | .90 | 1.36 | .37 | .56 |
| ammonia | .82 |  | .76 |  |
| hydrogen sulfide | .49 |  | .20 |  |
| water | .89 |  | .37 |  |

liquid product from the coker was mixed with a proportionate amount of fresh *in situ* crude shale oil. The mixture of crude shale oil and coker distillate was then fractionated to obtain fractions similar to those from the first distillation. Residuum from the second distillation was coked and liquid product from the coker again mixed with a proportionate amount of crude shale oil. The mixture of coker distillate and crude shale oil was fractionated to obtain a 350°F end-point naphtha, a 350°–550°F light oil, and a 550°–850°F heavy oil to be used as charge stocks for the preparation of synthetic crude oil. Yields from the first and third distillations and coking are compared in Table IV. Properties of the liquid products are shown in Table V.

The 550°–850°F heavy oil from the third distillation was hydrogenated in a continuous 174-hr run. No deactivation of the catalyst was evident from analyses of samples of the liquid product taken during the run, and yields of liquid and gaseous product and hydrogen consumption were constant with time throughout the total test. Operating conditions and yields of products are in Table VI. Approximately 40 wt % of the heavy oil was converted to products boiling below 550°F, of which

## Table VII.    Hydrogenation of 350°–550°F Light Oil

Operating conditions

| | |
|---|---|
| temperature (°F) | 750 |
| pressure (psig) | 1500 |
| space velocity ($W_o/W_c/hr$) | 1.0 |
| hydrogen feed (scf/bbl) | 5000 |
| hydrogen consumed (scf/bbl) | 1050 |

| | % of Feed | | % of Crude | |
|---|---|---|---|---|
| Yields of Products | Weight | Volume | Weight | Volume |
| $C_5+$ liquid product | 98.00 | 102.93 | 55.19 | 60.36 |
| $C_5$–175°F | 1.22 | 1.54 | .69 | .90 |
| 175°–350°F | 15.66 | 17.10 | 8.82 | 10.03 |
| 350°–550°F | 81.12 | 84.29 | 45.68 | 49.45 |
| coke | .19 | | .11 | |
| hydrogen | −1.78 | | −1.00 | |
| methane | .17 | | .10 | |
| ethane | .24 | | .13 | |
| propane | .20 | | .11 | |
| isobutane | .19 | .28 | .11 | .17 |
| butane | .56 | .82 | .32 | .49 |
| ammonia | 1.24 | | .70 | |
| hydrogen sulfide | .49 | | .28 | |
| water | .50 | | .28 | |

### Table VIII. Hydrogenation of 175°–350°F Naphtha

Operating conditions

| | |
|---|---|
| temperature (°F) | 700 |
| pressure (psig) | 1500 |
| space velocity ($V_o/V_c$/hr) | 1.0 |
| hydrogen feed (scf/bbl) | 5000 |
| hydrogen consumed (scf/bbl) | 720 |

| Yields of Products | % of Feed | | % of Crude | |
|---|---|---|---|---|
| | Weight | Volume | Weight | Volume |
| $C_5+$ liquid product | 99.83 | 102.09 | 20.45 | 23.67 |
| $C_5$–175°F | 4.14 | 4.40 | .85 | 1.02 |
| 175°–350°F | 95.69 | 97.69 | 19.60 | 22.65 |
| coke | .01 | | .00 | |
| hydrogen | −1.39 | | −.28 | |
| methane | .00 | | .00 | |
| ethane | .13 | | .02 | |
| propane | .24 | | .05 | |
| isobutane | .01 | .01 | .00 | — |
| butane | .02 | .03 | .00 | — |
| ammonia | .59 | | .12 | |
| hydrogen sulfide | .32 | | .07 | |
| water | .24 | | .05 | |

23.75% was light oil boiling between 350° and 550°F, and 9.25 wt % was material boiling between 170° and 350°F.

The 350°–550°F light oil obtained from hydrogenation of the 550°–850°F heavy oil was combined with the 350°–550°F light oil from the third distillation of the *in situ* crude shale oil and coker distillate. The combined light oils were hydrogenated in a continuous 75-hr run. Operating conditions and product yields are shown in Table VII. Approximately 39 wt % of the 350°–550°F material was converted to products boiling below 350°F, of which 15.7 wt % was naphtha boiling between 175° and 350°F.

The 175°–350°F naphthas from the two previous hydrogenation runs were combined with the total naphtha from the third distillation of *in situ* crude shale oil and coker distillate. The combined naphthas were then hydrogenated in a continuous 48-hr run. Operating conditions and product yields are shown in Table VIII. Under the conditions used, only 4.3 wt % of the charge was converted to products boiling below 175°F.

Properties of the charge stocks and liquid products from the various hydrogenation runs are shown in Table IX. Hydrogenation of the 550°–850°F heavy oil under the conditions used reduced the nitrogen content from 16,100 ppm in the charge stock to 935 ppm in the 550°–850°F

**Table IX.  Properties of Charge Stocks and Liquid Products from Hydrogenation Runs**

|  | 550°–850°F Heavy Oil | 350°–550°F Light Oil | 175°–350°F Naphtha |
|---|---|---|---|
| Charge stock |  |  |  |
| gravity (°API) | 27.2 | 35.0 | 47.3 |
| nitrogen (ppm) | 16,100 | 10,850 | 4,900 |
| sulfur (ppm) | 4,630 | 4,590 | 3,010 |
| Liquid product |  |  |  |
| gravity (°API) | 37.6 | 43.9 | 53.1 |
| nitrogen (ppm) | 880 | 70 | 1.3 |
| sulfur (ppm) | 9 | 7 | 16 |
| $C_5$–175°F naphtha |  |  |  |
| gravity (°API) | 77.2 | 77.6 | 60.8 |
| nitrogen (ppm) | 3.7 | 3.8 | .5 |
| sulfur (ppm) | 22 | 7 | .3 |
| 175°–350°F naphtha |  |  |  |
| gravity (°API) | 53.3 | 49.6 | 52.6 |
| nitrogen (ppm) | 299 | 53 | .8 |
| sulfur (ppm) | 9 | 3 | 10 |
| 350°–550°F light oil |  |  |  |
| gravity (°API) | 35.0 | 41.5 | — |
| nitrogen (ppm) | 1,220 | 79 | — |
| sulfur (ppm) | 8 | 1.2 | — |
| 550°–850°F heavy oil |  |  |  |
| gravity (°API) | 35.6 | — | — |
| nitrogen (ppm) | 935 | — | — |
| sulfur (ppm) | 9 | — | — |

fraction of the hydrogenated liquid product. Sulfur was reduced from 4630 to 9 ppm and the gravity increased from 27.2° to 35.6° API.

Hydrogenation of the 350°–550°F light oil under the conditions used reduced the nitrogen content from 10,850 ppm in the charge stock to 79 ppm in the 350°–550°F fraction of the hydrogenated liquid product. Sulfur was reduced from 4590 to 1.2 ppm, and the gravity increased from 35.0° to 41.5° API.

Hydrogenation of the 175°–350°F naphtha under the conditions used reduced the nitrogen content from 4900 ppm in the charge stock to less than 1 ppm in the 175°–350°F fraction of the hydrogenated liquid product. Sulfur was reduced from 3010 to 10 ppm, and the gravity increased from 47.3° to 52.6° API.

These results indicate that nitrogen removal is considerably more efficient when the naphtha, light oil, and heavy oil are hydrogenated separately. For example, the light oil produced during hydrogenation of the 550°–850°F heavy oil contained 1220 ppm nitrogen; the light oil produced by hydrogenation of the 350°–550°F light oil contained only 79 ppm nitrogen. The 175°–350°F naphtha produced during hydrogenation of the heavy oil contained 299 ppm nitrogen and that produced during hydrogenation of the light oil contained 53 ppm nitrogen, but the 175°–350°F naphtha produced during hydrogenation of the heavy oil contained 299 ppm nitrogen and that produced during hydrogenation of the light oil contained 53 ppm nitrogen. However the 175°–350°F naphtha produced during hydrogenation of the naphtha fraction contained only 0.8 ppm of nitrogen.

In Table X the properties of the syncrude prepared from *in situ* crude shale oil are compared with the properties of a syncrude listed by the NPC. Relative amounts and properties of the naphthas, light oils, and heavy oils are also compared. These data show that the nitrogen content, sulfur content, pour point, viscosity, and API gravity of syncrude prepared from *in situ* crude shale oil are lower than those suggested in

Table X. Comparison of NPC and *in Situ* Syncrudes and Distillate Fractions

| | NPC | in Situ |
|---|---|---|
| Syncrude | | |
| gravity (°API) | 46.2 | 43.9 |
| pour point (°F) | 50 | <32 |
| viscosity (SUS at 100°F) | 40 | 32 |
| nitrogen (ppm) | 350 | 250 |
| sulfur (ppm) | 50 | 5 |
| Butanes and butenes (vol %) | 9.0 | 1.7 |
| C₅–350°F naphtha (vol %) | 27.5 | 24.8 |
| gravity (°API) | 54.5 | 54.7 |
| nitrogen (ppm) | 1 | 1 |
| sulfur (ppm) | <1 | 8 |
| aromatics (vol %) | 18 | 14 |
| naphthenes (vol %) | 37 | 44 |
| paraffins (vol %) | 45 | 42 |
| 350°–550°F distillate (vol %) | 41.0 | 48.1 |
| gravity (°API) | 38.3 | 41.5 |
| nitrogen (ppm) | 75 | 79 |
| sulfur (ppm) | 8 | 1.2 |
| aromatics (vol %) | 34 | 24 |
| freezing point (°F) | −35 | −29 |
| 550°–850°F distillate (vol %) | 22.5 | 25.4 |
| gravity (°API) | 33.1 | 35.6 |
| nitrogen (ppm) | 1,200 | 935 |
| sulfur (ppm) | <100 | 9 |
| pour point (°F) | 80 | 55 |

Table XI.    Summary of Yields from Preparation of Syncrude

| Process[a] | Wt % | | | | |
| | A | B | C | D | Totals |
|---|---|---|---|---|---|
| $C_5$–175°F naphtha | — | 0.80 | 0.69 | 0.85 | 2.34 |
| 175°–350° naphtha | — | — | — | 19.60 | 19.60 |
| 350°–550°F light oil | — | — | 45.68 | — | 45.68 |
| 550°–850° heavy oil | — | 25.04 | — | — | 25.04 |
| Coke | 3.08 | .05 | .11 | — | 3.24 |
| Hydrogen | 3.02 | −.93 | −1.00 | −.28 | −2.19 |
| Methane | .27 | .32 | .10 | — | .69 |
| Ethane | .16 | .36 | .13 | .03 | .68 |
| Ethylene | .03 | — | — | — | .03 |
| Propane | .20 | .37 | .11 | .04 | .72 |
| Propylene | .09 | — | — | — | .09 |
| Isobutane | .03 | .12 | .11 | — | .26 |
| Butane | .10 | .37 | .32 | — | .79 |
| Butenes | .10 | — | — | — | .10 |
| Carbon monoxide | .03 | — | — | — | .03 |
| Carbon dioxide | .02 | — | — | — | .02 |
| Ammonia | — | .76 | .70 | .11 | 1.57 |
| Hydrogen sulfide | .07 | .20 | .28 | .07 | .62 |
| Water | — | .37 | .28 | .04 | .69 |

[a] A. Distillation and coking. B. Hydrogenation of 550°–850°F heavy oil. C. Hydrogenation of 350°–550°F light oil. D. Hydrogenation of 175°–350°F naphtha.

the NPC report. The lower gravity of syncrude prepared from *in situ* crude shale oil may be attributable in part to the lower content of butanes and butenes and in part to the greater volumes of materials boiling above the naphtha range. The sulfur content of the naphtha is somewhat high, but the sulfur contents of the other fractions are much lower than those suggested by the NPC.

A summary of the yields from the various steps used in the preparation of syncrude from *in situ* crude shale oil is shown in Table XI. The overall yield of syncrude was 103 vol % of the original crude.

### Summary and Conclusions

Hydrogenation tests made on the 600°–1000°F heavy gas oil from *in situ* crude shale oil showed that a nickel–molybdenum-on-alumina catalyst was superior to either cobalt–molybdenum-on-alumina or nickel–tungsten-on-alumina catalysts for removing nitrogen from shale oil fractions. This nickel–molybdenum-on-alumina catalyst was used in the preparation of the synthetic crude oil. A high yield of premium refinery feedstock whose properties compared favorably with those of a syncrude described by the NPC was attained by hydrogenating the naphtha, light oil, and heavy oil fractions obtained from *in situ* crude shale oil by distillation and coking of the vacuum residues. The overall yield of syncrude was 103 vol % of the original crude shale oil and contained only 250 ppm nitrogen and 5 ppm sulfur. It had a pour point below 32°F and a viscosity of 32 SUS at 100°F.

*Literature Cited*

1. Burwell, E. L., Carpenter, H. C., Sohns, H. W., "Experimental *In Situ* Retorting of Oil Shale at Rock Springs, Wyo.," *U.S. Bur. Mines Tech. Prog. Rep.* (1969) **16**, 8.
2. Burwell, E. L., Sterner, T. E., Carpenter, H. C., "Shale Oil Recovery by *In Situ* Retorting—A Pilot Study," *J. Petrol. Technol.* (1970) 1520.
3. Carpenter, H. C., Burwell, E. L., Sohns, H. W., "Exaluation of an *In Situ* Retorting Experiment in Green River Oil Shale," *J. Petrol. Technol.* (1972) 21.
4. Harak, A. E., Dockter, L., Carpenter, H. C., "Some Results from the Operation of a 150-Ton Oil Shale Retort," *U.S. Bur. Mines Tech. Prog. Rep.* (1971) **30**, 14.
5. Carpenter, H. C., Tihen, S. S., Sohns, H. W., "Retorting Ungraded Oil Shale as Related to *In Situ* Processing," *Am. Chem. Soc., Div. Petrol. Chem., Preprint* **13** (2) F50 (April, 1968).
6. Dockter, L., Long, Jr., A., Harak, A. E., "Retorting Ungraded Oil Shale as Related to *In Situ* Processing," *Am. Chem. Soc., Div. Fuel Chem., Preprints* **15** (1) 2 (1971).
7. Jensen, H. B., Poulson, R. E., Cook, G. L., "Characterization of a Shale Oil Produced by *In Situ* Retorting," *Am. Chem. Soc., Div. Fuel Chem., Preprints* **15** (1) 113 (1971).
8. Harak, A. E., Long, Jr., A., Carpenter, H. C., "Preliminary Design and Operation of a 150-Ton Oil Shale Retort," *Q. Colo. Sch. Mines* (1970) **65** (4) 41.
9. "U.S. Energy Outlook, An Interim Report," National Petroleum Council, **2**, 80, 1972.
10. Carpenter, H. C., Cottingham, P. L., "Evaluation of Catalysts for Hydrogenating Shale Oil," *U.S. Bur. Mines Rep. Invest.* **5533** (1959) 27.
11. Benson, D. B., Berg, L., "Catalytic Hydrotreating of Shale Oil," *Chem. Eng. Prog.* (1966) **62** (8) 61.
12. Montgomery, D. P., "Refining of Pyrolytic Shale Oil," *Ind. Eng. Chem. Prod. Res. Develop.* (1968) **7** (4) 274.

RECEIVED December 17, 1974. This work was supported by a cooperative agreement between the U.S. Energy Research and Development Administration, Laramie Energy Research Center, and the University of Wyoming. Reference to specific trade names or manufacturers does not imply endorsement by the U.S.E.R.D.A. Prior to completion of this work Laramie Energy Research Center was a unit of the U.S. Bureau of Mines.

# 7

# Solution of Silica in Green River Oil Shale

W. C. MEYER and T. F. YEN

University of Southern California, Los Angeles, Calif. 90007

*Bioleaching of Green River oil shale leaves a silica-enriched residue in which potentially petroliferous organics (primarily kerogen) remained trapped. Organic solvent systems using pyridine, quinoline or water, and potassium hydroxide were examined to ascertain their effectiveness in dissolving silica (crushed quartz $< 125\mu$), thereby liberating trapped organics. Solvent capability of each system was determined as percent weight loss from the initial silica sample. Quinoline plus potassium hydroxide pellets proved most effective, causing approximately 50% weight loss in 5 hr. Saturated aqueous potassium hydroxide yielded a weight loss of approximately 9%. The organic solvents appear to serve as a substrate to allow basic aqueous solution of silica at elevated temperatures.*

Petroliferous organic compounds contained in Green River oil shale can be released only after the entrapping mineral matrix, predominantly dolomite ($CaMg(CO_3)_2$), calcite ($CaCO_3$), and quartz ($SiO_2$) is broken down (*1*). Carbonates are readily solubilized in an acid medium (*1*), but removal of quartz presents a problem. On a previous occasion, potassium hydroxide (KOH) pellets in boiling pyridine dissolved through a borosilicate glass flask during distillation, suggesting that basic organic solvents with potassium hydroxide may be good silica solvents.

### Method

Experiments were designed to test the solubility of quartz in several solvent systems, prepared by the following method.

1. 200 ml quinoline plus 60 g potassium hydroxide pellets added directly.

2. 100 ml quinoline plus 100 ml saturated aqueous potassium hydroxide.

3. 200 ml pyridine plus 60 g potassium hydroxide pellets added directly.

4. 100 ml pyridine plus 100 ml saturated aqueous potassium hydroxide.

5. 200 ml glycerol plus 60 g potassium hydroxide pellets added directly.

6. 200 ml saturated aqueous potassium hydroxide.

To each of these systems a measured amount ($\cong 5$ g) of crushed ($< 125\,\mu$) quartz was added in a stainless-steel beaker, and the systems were heated to their boiling points at atmospheric pressure for 5 hr. Quartz was chosen for these experiments because of its tight crystalline structure and relative insolubility in relation to other silica species (2). Solubility values for this mineral would represent a minimum for a given solvent system with respect to silica.

### Table I.  Solvent Systems

| Solvent System | Temp (°C) | Time (hr) | Wt. Loss(%)[a] |
|---|---|---|---|
| *Quartz* | | | |
| Quinoline + KOH pellets | 235 | 5 | 55 |
| Quinoline + saturated aqueous KOH | 155 | 5 | 36 |
| Pyridine + KOH pellets | 130 | 5 | 3 |
| Pyridine + saturated aqueous KOH | 115 | 5 | 7 |
| Glycerol + KOH pellets | 179 | 5 | 29 |
| Saturated aqueous KOH | 110 | 5 | 9 |
| *Oil Shale* | | | |
| Quinoline + KOH pellets | 235 | 5 | 37 |

[a] Figures are mean values for duplicate runs.

Solvent vapor was refluxed into each system by placing a volumetric flask of cold water over the mouth of the beaker. Undissolved quartz was trapped in base-resistant sharkskin filter paper and thoroughly washed with distilled water. The residue was then dried and weighed. Weight loss is expressed as percent of original weight (Table I).

Problems were encountered in cleaning the undissolved quartz of gummy residue and precipitated potassium hydroxide which often formed as the solution cooled during filtering. In most cases, the residue was water soluble and was eliminated by repeated washing.

### Discussion

The accumulated data (Table I) showed that, of the solvent systems investigated, quinoline plus potassium hydroxide pellets provided the best results, followed by a mixture of quinoline and aqueous saturated potas-

Quartz + Water

$$SiO_2 + 2H_2O \rightsquigarrow H_4SiO_4 \rightsquigarrow 3H^+ + HSiO_4^{-3}$$

by

$$H_4SiO_4 \rightsquigarrow H^+ + H_3SiO_4^-$$

$$H_3SiO_4^- \rightsquigarrow H^+ + H_2SiO_4^{-2}$$

$$H_2SiO_4^{-2} \rightsquigarrow H^+ + HSiO_4^{-3}$$

Add Base

$$H^+ + OH^- \rightsquigarrow H_2O \ . \ \dot{} \ .$$

$$SiO_2 + 2H_2O + OH^- \rightsquigarrow H_4SiO_4 + OH^- \rightsquigarrow 3H_2O + HSiO_4^{-3}$$

$$OH^- = H^+ \text{ sink}$$

Equilibrium is unbalanced and goes to the right

*Figure 1.   Reaction of silica with water as basic medium*

sium hydroxide. Mixtures of pyridine and potassium hydroxide were not as effective.

It is not clear why a basic quinoline system dissolves more quartz than a basic aqueous system, especially since potassium hydroxide is not readily soluble in quinoline. Attempts to dissolve potassium hydroxide in quinoline result in a dense milky suspension. The suspension may contain the active agent, perhaps a quaternary organic salt, that serves as a better hydroxyl source than solid potassium hydroxide in water. Practical grade quinoline was used for these experiments, and this may contain enough water for dissociation.

A more probable explanation is that all the studied systems are undergoing aqueous potassium hydroxide reaction with quartz (Figure 1) and that the different degrees of solvent efficiency are merely a direct function of the temperature at which these reactions occur. Siever (3) found that the solubility of silica in water increases with temperature, and it seems reasonable to assume that this general relationship is true for aqueous potassium hydroxide systems as well. All solvents used for these experiments, with the exception of pyridine, were practical grade and would be expected to contain small, but perhaps significant, amounts of water. There seems to be a general correlation between solvent capability and boiling point of the system (Figure 2), suggesting that the organic solvents serve merely as a substrate to elevate the temperature of aqueous reactions.

A glycerol system was chosen to test this hypothesis, because the boiling point of this chemical is considerably higher than that of other solvents used in these experiments. It was found that glycerol and potassium hydroxide form a complex with a relatively low boiling point of 179°C, nonetheless the temperature and solvent capabilities of this system are in keeping with the proposed thesis (Figure 2). The pyridine system produced surprisingly poor results, but since reagent grade pyridine should have a negligible water content, the possibility of a good aqueous reaction is eliminated.

Addition of aqueous potassium hydroxide solution to the organic solvent would be expected to provide ample water for reaction but would also decrease the boiling point of the system. The decrease in solvent efficiency of those systems that were aqueous by design might then be attributed to the decreased reaction temperature. To support or disprove this hypothesis, additional experiments were made to ascertain the solvent capability of the studied systems at the same temperature.

Four solvent systems—potassium hydroxide plus glycerol, saturated aqueous potassium hydroxide, potassium hydroxide plus pyridine, and potassium hydroxide plus quinoline—were prepared as previously described. All four systems were allowed to react with quartz for 5 hr at

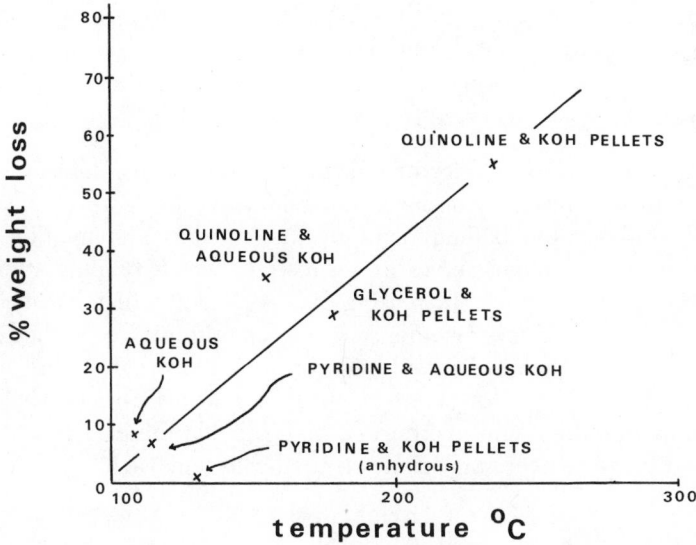

*Figure 2. Relationship of solvent capability to temperature (boiling point of system)*

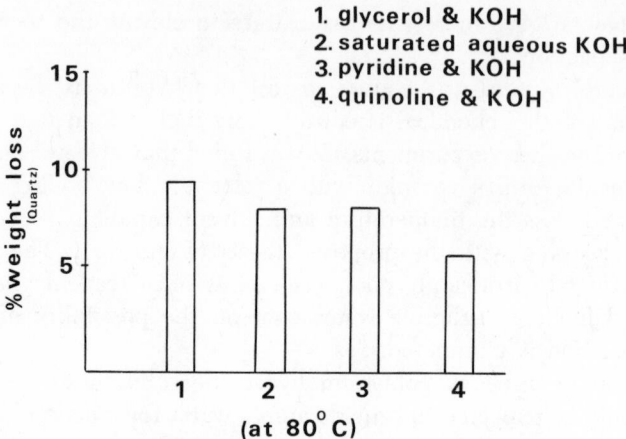

*Figure 3. Relationship of solvent effectiveness
of different systems at the same temperature.
All experiments run for 5 hr.*

80°C. The results (Figure 3) show no significant difference in solvent capability at constant temperature, further indicating that temperature and not solvent is the significant variable.

To test the practical applicability of the quinoline solvent systems, samples of raw, crushed ($< 125\,\mu$) Green River oil shale were extracted in the previously described manner (Table I). The samples lost 37% of their weight, suggesting that a large portion of the quartz and other silicates had been dissolved from the rock.

### Conclusions

The increased solubility of quartz in basic organic solvent systems appears to be caused by aqueous potassium hydroxide reaction at temperatures above the boiling point of the aqueous system alone. The organic solvent fraction serves as a substrate which permits attainment of elevated temperatures. Increasing the pressure at which basic aqueous reactions are performed would serve as an alternative method which would eliminate the need for addition of organic solvents. This prospect is especially attractive for *in situ* removal of silicates from oil shale since geothermal gradient and overburden may provide the elevated temperature and pressure necessary for efficient silicate removal.

### Literature Cited

1. Meyer, W. C., Yen, T. F., "The Effects of Bioleaching in Green River Oil Shale," *Am. Chem. Soc., Div. Fuel Chem., Preprint* (1974) **19,** 94.

2. Iler, R. K., "The Colloid Chemistry of Silica and Silicates," p. 6, Cornell Univ., New York, 1955.
3. Siever, R., "Silica Solubility 0–200°C, and the Diagenesis of Silican Sediments," *J. Geol.* (1962) **70**, 127.

RECEIVED December 17, 1974. This work is supported by NSF Grant No. GI-35683, AER-74-23797, and A.G.A. Grant No. GR-48-12.

# 8

# Fracturing Oil Shale with Explosives for *in Situ* Recovery

J. S. MILLER and ROBERT T. JOHANSEN

Bartlesville Energy Research Center, Bartlesville, Okla. 74003

*Three different explosive fracturing techniques were developed by the Bureau of Mines for preparing oil shale for* in situ *recovery near Rock Springs, Wyo. (1) Displacing and detonating nitroglycerin in natural or hydraulically induced fracture systems; sufficient fragmentation was obtained to sustain an* in situ *combustion experiment. (2) Displacing and detonating nitroglycerin in induced fractures followed by wellbore shots using pelletized TNT; suitable interwell fragmentation was achieved for* in situ *recovery experimentation. (3) Detonating wellbore charges using pelletized TNT; suitable interwell fragmentation was not achieved to support an* in situ *recovery experiment.*

The research described here was started in 1964 as a part of the energy research program of the Bureau of Mines to develop methods to fragment the oil shale with explosives and to expose sufficient rock surface area to achieve *in situ* combustion recovery of shale oil. These studies are relevant to the rising concern for the capability of the U.S. to meet its mounting energy demands at reasonable costs and an acceptable level of social and environmental impact.

The concept involves the injection and detonation of a liquid chemical explosive in natural or previously induced fracture systems or the use of a pelletized explosive to enlarge and extend these fractures to provide fragmentation and interwell communication. This study is one of few known research efforts to evaluate results of detonating sheetlike layers of explosive to increase flow capacity in confined rock formations. The literature contained little information to guide the design of the experiments. Some related work, however, had been conducted by a few individuals and oil field service companies. Briefly, the earlier work resulted in moderate successes, near failures, injuries, and numerous premature detonations that destroyed wells and property. One source (1) reported a combined shot of 5000-qt nitroglycerin (NGl) which was displaced into the formation from a wellbore loaded with glass marbles in the Turner Valley field during February, 1946. Oil flow was not increased, and no further shooting was done.

Brewer (2) indicated that the Tar Springs, Jackson, and Benoist Formations in the Illinois Basin responded when the voids in these low permeability formations were filled with explosives and detonated. Further, the Cleveland and Red Fork sands in Oklahoma were reported to have responded to NGl shots in the formation. Data on individual tests and detailed results were not publicized.

Zandmer (3, 4), Brandon (5), Hanson (6), and Hinson (7) filed patents relating to explosive fracturing. Results of the patented Stratablast process were reported at a meeting of the American Petroleum Institute in April, 1965 (8). The multiple component systems used were generally hypergolic fluids that explode when combined in the formation. In 1970, an article (9) reviewed the "new look" at stimulation by explosives and gave a state-of-the-art account of the modern explosive techniques for improving production.

The extensive oil shale formations in parts of Wyoming, Colorado, and Utah cover an area of approximately 16,000 sq mi (Figure 1). These rocks of the Green River Formation originated as limey muds deposited in predominantly lacustrine environments. Through geologic processes these lake floor deposits were transformed into marlstone containing organic kerogen, which requires considerable heat to change it into a liquid shale oil. Because the host rock has little natural porosity and permeability, fractures must be induced through which air can be injected to establish and maintain a combustion zone and to provide a means to recover the retorted shale oil.

Surveys (10) show that oil shale deposits in the U.S. testing 25 gal or more per ton contain about 600 billion bbl of oil. These deposits range in depths from surface outcrops to 2000 ft. If a lower limit of richness is set at 10 gal/ton, the available volume of oil would be increased 25-fold to about 2 trillion bbl. The development of a technique for efficient shale oil recovery would significantly influence the U.S.'s total oil supply.

Development of mining and aboveground retorting of oil shale has only recently advanced beyond the experimental stage. In addition, aboveground oil shale processing is accompanied by ecologic disturbances with the attendant water supply and pollution problems of effluents and disposal of spent shale. Underground retorting potentially offers a more

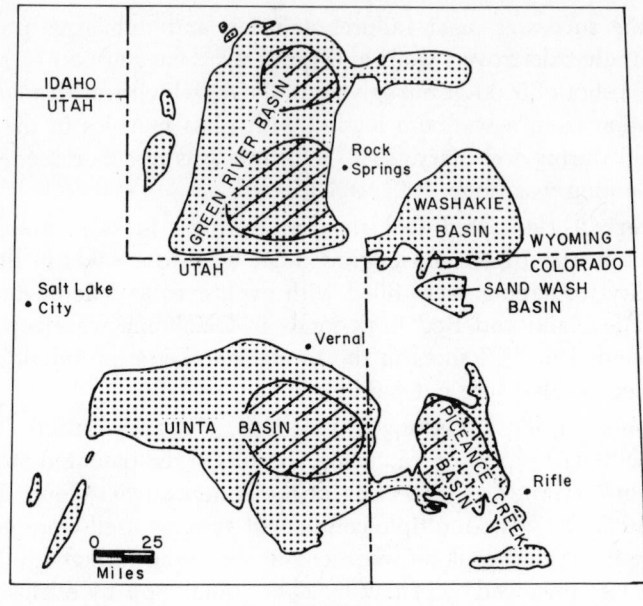

LEGEND

Area of 25 gal/ton, or richer, oil shale more than 10 ft thick        Area of Green River Formation

*Figure 1.   Location of oil shale deposits in Utah, Colorado, and Wyoming*

feasible solution to the problem. Bureau of Mines laboratory and field research on the use of chemical explosives to fracture the rock lends encouragement for developing means to accommodate the airflow requirements to maintain combustion and for displacing retorted shale oil to producing wells.

Explosive fracturing was applied to the Green River Formation on eight sites near Rock Springs and Green River, Wyo., three of which are described in this paper. Descriptions of procedures and explosive fracturing evaluation methods relating to the sites have been reported (*11, 12, 13*). The methods used to fragment the formation differed from site to site because of the differences in depth to the richer shale beds at the various sites, differences of ground water levels, the extent of natural or induced fractures encountered, and the type of explosive used to fracture the shale.

### Field Test, Rock Springs Site 4

**Purpose.** The first research program designed to recover shale oil by *in situ* combustion was planned for Rock Springs site 4. Little information has been published about *in situ* retorting methods for production

of shale oil ( *14, 15, 16* ). The experiment was designed to establish sufficient fracture permeability through expanding natural fractures, inducing
hydraulic fractures, and fracturing using chemical explosives ( *17* ).

**Procedure.** The site was developed on a five-spot pattern about
25 ft square, as shown in Figure 2. The wells were rotary drilled with
water and completed with 50 ft of 7-in. casing and cemented to the surface. A 6¼-in. hole was drilled below the casing to a total depth of
100 ft in the oil shale. Two additional wells were drilled off pattern as
observation wells. A Fischer assay determined the oil yield of the section
as 19.0–26.5 gal/ton. Two sand-propped hydraulic fracture treatments
were applied for emplacing NGl in the formation.

In the first two tests of a series of three explosive fracturing experiments, well 3 was used to inject and displace 100 and 300 qt of NGl in
the depth intervals from 70 to 74 ft. Continuous sampling of surrounding

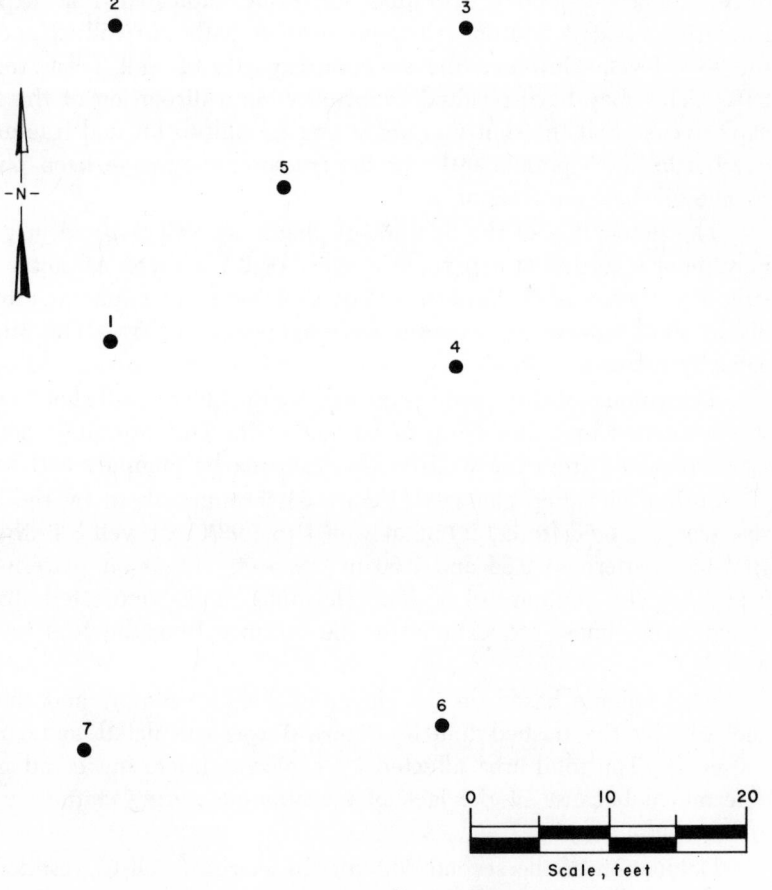

*Figure 2.    Location of wells, Rock Springs site 4*

test wells showed that the NGl migrated to a second well during each injection. Detonators were set in each well, and the explosive charges were detonated simultaneously.

To improve interwell communication further, a hydraulic fracturing treatment was performed at a depth of 79 to 84 ft in well 5 to ensure the displacement and detonation of the 300-qt charge of NGl. The effectiveness of the three fracturing techniques was determined by measuring airflow rates between selected wells before and after each test.

**Results.** The first explosive fracturing test detonated 100 qt of NGl displaced into the formation from well 3 at a depth interval from 70 to 74 ft. Following detonations in wells 3 and 4, fracture intervals in the wellbores connecting the injection well 3 and other wells were determined by airflow measuerments.

Comparing these airflow intervals with those permeable zones induced by conventional hydraulic fracturing indicated that explosive fracturing created additional communication paths to wells 2 and 5 at the 73-ft level. However, the injection capacity of well 3 was reduced 64%. This may have resulted from too wide a dispersion of the liquid explosive, so that the shot was not strong enough to lift and fracture the overburden rock permanently, or the fractures may have been plugged by fine oil-shale particles or mud.

The detonation of the first 300-qt charge of NGl resulted in ground movement recorded at a particle velocity of 2.5 in./sec. Air-entry intervals that existed after the first 100-qt shot were not apparent after the 300-qt shot; however, new zones were opened to airflow. The injection capacity increased 500%.

The volume of the fractures created by the 300-qt NGl shot in well 3 was estimated by water fillup to be 800 cu ft. This was the amount of water removed from the wells in the test area by pumping and bailing.

Surface-elevation changes (Figure 3), brought about by the explosive work, ranged from 1.20 in. at well 1 to 1.92 in. at well 3 in the five-spot test pattern to 0.84 and 0.60 in., respectively, at off pattern wells 6 and 7. The contours of surface elevation change indicated that the change was almost proportional to the distance from the NGl injection well 3.

Void volume based on the elevation-change contours and the area enclosed by the dashed line in Figure 3 was calculated to be nearly 150 cu ft. The total area affected by explosive fracturing could not be determined because of the lack of elevation-measuring stations outside of the contoured area.

Detonation of the second 300-qt NGl charge (well 5) resulted in a particle velocity of 2.2 in./sec measured at the surface, indicating com-

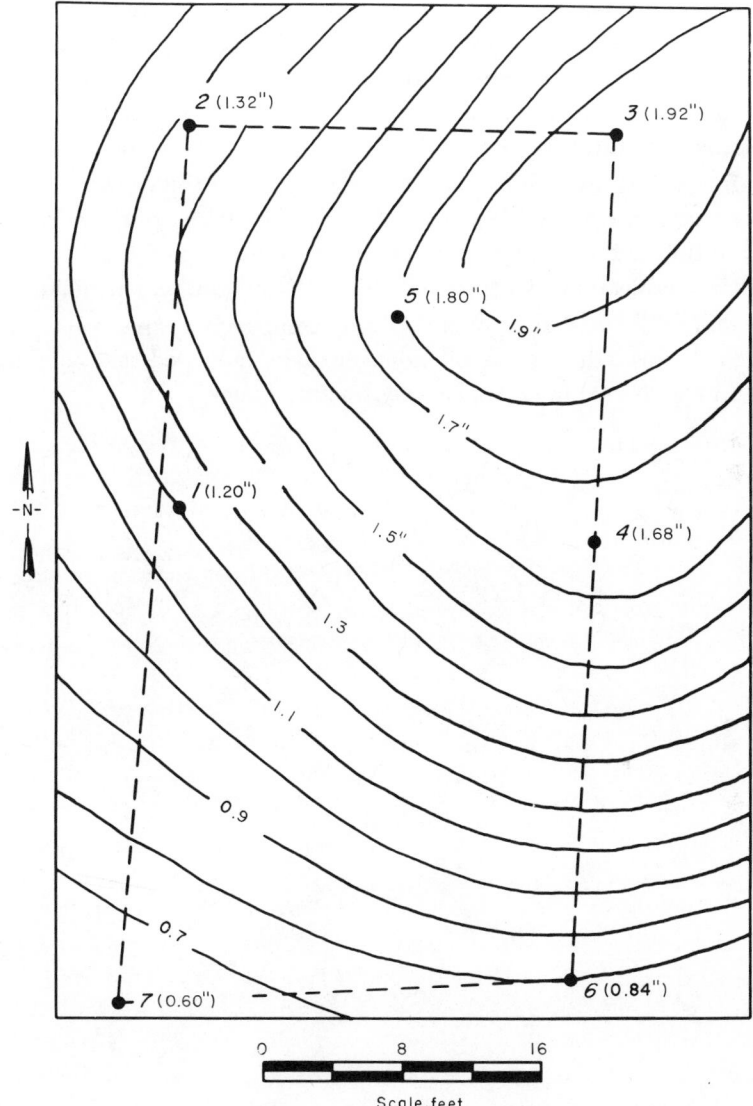

*Figure 3. Contours of surface elevation change resulting from 300-qt. NGl shot, Rock Springs site 4*

plete detonation. Airflow tests were made, and the air-injection capacity increased about 800%. These air-injection rates were judged to be suffi-cient to support a planned *in situ* combustion experiment.

Although the nature and extent of fractures created in the oil shale by the various fracturing techniques are not completely known, some generalizations can be made. Horizontal fractures were opened to all wells in the original five-spot pattern with no apparent vertical communication established, except in the area between wells 3 and 5 where greater rock breakage with horizontal and vertical fracturing resulted from the explosive fracturing.

In general, hydraulic fracturing with sand propping provided adequate void space for emplacement of the NGl in these explosive-fracturing tests in the oil shale. Explosive fracturing caused significant increases in fracture permeability when a sufficient NGl charge was detonated.

Results from a subsequent *in situ* combustion experiment on this site (*18*), to produce shale oil from oil shale, indicated that combustion could be sustained in an explosively fractured zone.

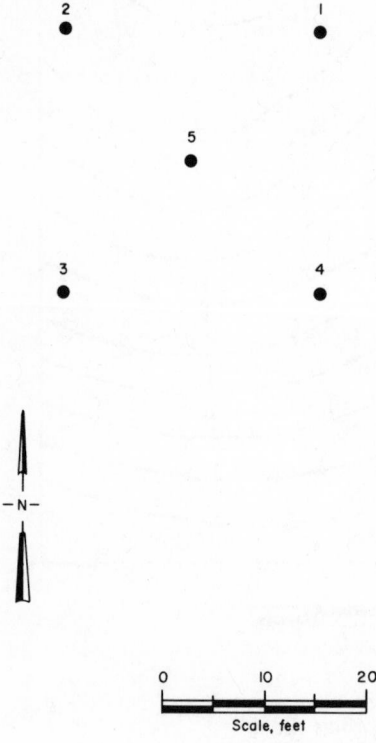

Figure 4.  Location of wells, Rock Springs site 5

## Field Test, Rock Springs Site 5

**Purpose.** Explosive-fracturing research at Rock Springs site 5 was designed to develop additional expertise in creating sufficient fragmentation and permeability in the oil shale to support *in situ* retorting. Results obtained from previously completed field applications indicated that detonation of a liquid explosive in natural or hydraulic fractures effectively lifted the overburden, extended existing fractures, and fragmented the oil shale formations (*11, 19*). At this stage of the research, it was not possible to either precisely describe or adequately evaluate the fractures. Consequently, to achieve maximum fracturing, a method of explosive fracturing was used which combined displacement and detonation of a liquid chemical explosive in a natural fracture system and use of pelletized TNT in a series of wellbore shots as the principal means to fragment the oil shale for *in situ* retorting.

**Procedure.** SITE PREPARATION. A five-spot pattern of test wells (Figure 4) was drilled to approximately 57 ft and completed with 7-in. casing cemented to the surface. The wells, deepened to 100 ft with a 6¼-in. bit, were tested to determine the extent of air communication between the center well (well 5) and the surrounding wells. These tests indicated fractures 1–3 ft in height between depths from 67 to 90 ft.

EXPLOSIVE FRACTURING. To fragment the oil shale, three types of explosives were used: desensitized NGl, 60-% dynamite, and pelletized TNT. Figure 5 shows the positions and sequence of all shots on Rock Springs site 5. A 340-qt charge of NGl was displaced from well 5 into the natural vertical fracture system (shot A, Figure 5) and was detonated successfully. This detonation was intended to lift the overburden and create space for fragmenting more shale through repetitive simultaneous wellbore shooting with other explosives. Elevation measurements were obtained on the casing heads of each well before and after detonation to determine residual crowning of the overburden rock.

During the second step of the fracturing experiments at this site, 60% dynamite was detonated in the five wells to relieve stress conditions in the block of oil shale. Each of the wells in the 25- by 25-ft five-spot pattern was loaded with 45-lb charges of 60% dynamite on detonating cord with electric caps attached and detonated simultaneously (shot B).

Theoretically, to fragment the block of oil shale by detonating wellbore charges of pelletized TNT, the area around center well 5 should be enlarged or "sprung." This would be accomplished by repeated wellbore shots from bottom to top of the test zone. The broken and enlarged area surrounding the wellbore would serve as a free face to enhance effects from later simultaneous wellbore shots across the pattern.

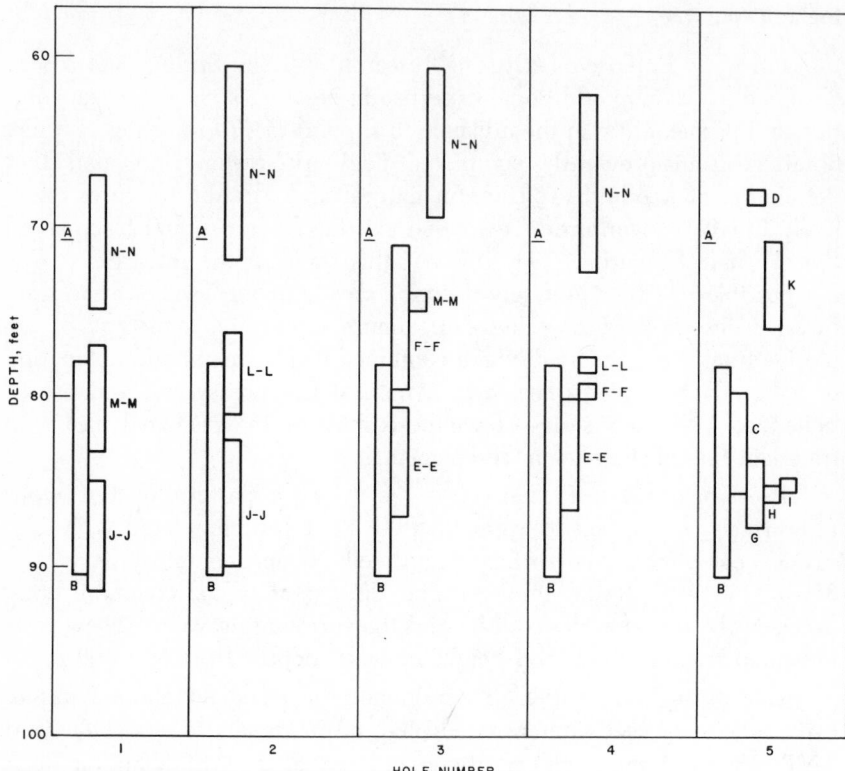

*Figure 5. Position of explosive charge in sequence of wellbore shots in five-spot pattern, Rock Springs site 5*

Six shots (C, D, G, H, I, K) using approximately 1000 lb of TNT, were detonated in well 5 at depths ranging from 67 to 88 ft. The first three shots were not stemmed; consequently, water and debris were blown to the atmosphere. The last three shots were sand tamped to the surface to fragment the maximum amount of oil shale around the wellbore and to permit the contained explosive gases to extend the induced fractures.

Two shots (E–E, F–F) using a total of 536 lbs of TNT were detonated in wells 3 and 4 between depths of 71 and 87 ft. After cleanout in wells 1 and 2, 150-lb charges of TNT were placed in each hole to depths of 85 and 83 ft, respectively, and detonated (shot J–J). Wells 2 and 4 were cleaned out, and a total of 250 lb of TNT filled the holes to depths of 76 and 77 ft, respectively, and were detonated (shot L–L). Wells 1 and 3 were prepared for reshooting by charging 150 lb of TNT in each wellbore at depths of 77 and 74 ft, respectively, and were detonated (shot M–M). This explosive fracturing series was concluded by loading the four outside wells—1, 2, 3, and 4—with charges of 296, 225, 185, and 185 lb of TNT and shooting simultaneously at depths of 75, 72, 69, and 73 ft, respectively (shot N–N).

**Results.** Although the numerous methods used to evaluate underground fractures created by confined explosive fracturing techniques in oil shale under this site revealed much information, the data obtained from the evaluation tests showed that the oil shale formation exposed to the effects of explosive fracturing was extensively fragmented. The data also indicated that the fragmented zone was roughly ovaloidal in shape, approximately 95 ft in diameter and 70 ft thick. Extensive fracture systems were detected by airflow tests 90 ft from the center well of the five-spot pattern.

*Field Test, Green River Site 1*

**Purpose.** This fracturing program was intended to devise an effective method to fracture the formation with wellbore shots. More specifically, Green River site 1 was developed to test chemical explosive fracturing procedures for establishing communication between wells at greater depths and well spacings than had been previously attempted in oil shale.

**Procedure.** Green River site 1 was located 5 mi west of the Rock Springs sites 4 and 5. The oil shale zone of interest, at approximately 340–385 ft, was selected after studying the analysis of cores cut from an earlier well. As determined by Fischer assay, oil yield of the cored section averaged about 21.0 gal/ton.

The completed site contained 10 pattern wells for explosive-fracturing research. Six wells were drilled on 50-ft spacing to form a rectangle with three wells on a side. The remaining four wells were drilled on 25-ft spacing to form a five-spot pattern (Figure 6). All wells were completed similarly to the earlier described test wells.

Caliper and gamma ray logs were run to detect caving and borehole irregularities and to correlate the oil shale formations. Airflow tests were made to measure initial communication between injection well 6 and the remaining wells in the pattern.

The first series of explosive tests on the site was performed on the wells in the five-spot pattern. The amount of TNT used in each well was calculated from total-depth and caliper-log measurements. The accumulated water was bailed prior to lowering the priming devices and filling the wellbores with TNT.

The pelletized TNT was poured slowly into the wells until the TNT column rose above the water level to assure that the TNT had not

*Figure 6.   Location of wells, Green River site 1*

**Table I.   Shooting Data, First Shot, Small Five-Spot Pattern, Green River Site 1**

| Well No. | Total Depth of Well (ft) | Total Depth of Casing (ft) | Explosive Height (ft) | Sand Tamp Depth (ft) | Explosive Used (lb) |
|---|---|---|---|---|---|
| 2 | 389.5 | 342.0 | 355.0[a] | 150 | 420 |
| 3 | 374.5 | 343.0 | 347.0 | 150 | 600[b] |
| 4 | 386.5 | 341.0 | 345.0 | 150 | 1,020 |
| 5 | 382.5 | 341.0 | 327.0[c] | 150 | 600[b] |
| 6 | 383.5 | 342.0 | 355.0 | 150 | 900[b] |

[a] Explosive bridged in casing.
[b] Explosive did not detonate.
[c] Explosive bridged; washed out; no additional explosive added.

**Table II.    Shooting Data, First Shot, Large Pattern, Green River Site 1**

| Well No. | Total Depth of Well (ft) | Total Depth of Casing (ft) | Explosive Depth (ft) | Sand Tamp Depth (ft) | Explosive Used (lb) |
|---|---|---|---|---|---|
| 7 | 373.5 | 341.0 | 346.0 | 150 | 600 |
| 8 | 375.0 | 337.0 | 342.0 | 150 | 600 |
| 9 | 384.5 | 337.0 | 342.0 | 150 | 840 |
| 10 | 373.5 | 340.0 | 345.0 | 150 | 600[a] |
| 11 | 403.5 | 342.0 | 308.0 | 150 | 960[b] |

[a] Explosive did not detonate.
[b] Explosive in casing. Casing damaged; no clean out.

bridged in the well. The wells were loaded with a total of 3540 lb of TNT, as indicated in Table I. Each well was sand tamped from the top of the explosive to approximately 150 ft in the casing before the explosive was detonated.

An explosive-fracturing test was then performed on the remaining five wells in the large pattern using a total charge of 3600 lb of TNT, as indicated in Table II. The five-spot pattern wells were cleaned to bottom to remove rubble from the wellbores. During the cleanout, it became evident from the recovery of debris that the explosive had not detonated in wells 3, 5, and 6. Caliper logs run in these wells verified the findings.

Wells 3, 5, and 6 were reloaded with 1860 lb of TNT and shot (Table III). After cleanout and after caliper logs were obtained, airflow tests were run on the five-spot pattern wells to determine the relative extent of fragmentation and improvement in communication between wells.

The final explosive-fracturing test performed on this site was a simultaneous shot detonated in the wells of the five-spot pattern. Wire-line measurements were obtained to determine total depth, and caliper logs were run to determine wellbore enlargement from which to calculate the amount of TNT to fill each well. Water was swabbed and bailed

**Table III.    Shooting Data, Second Shot, Small Five-Spot Pattern, Green River Site 1**

| Well No. | Total Depth of Well (ft) | Total Depth of Casing (ft) | Explosive Depth (ft) | Sand Tamp Depth (ft) | Explosive Used (lb) |
|---|---|---|---|---|---|
| 2 | — | — | — | 150 | —[a] |
| 3 | 379.0 | 343.0 | 345.0 | 150 | 600 |
| 4 | — | — | — | 150 | —[b] |
| 5 | 382.0 | 341.0 | 346.0 | 150 | 540 |
| 6 | 381.0 | 342.0 | 346.0 | 150 | 720 |

[a] Set bridge plug at 290.0 ft.
[b] Set bridge plug at 337.0 ft.

from each well, the primers were run to total depth, and a predetermined amount of TNT was poured in each well. A total charge of 7140 lb of TNT was loaded in these wells and detonated (Table IV).

**Results.** Data obtained from evaluation tests indicated that the oil shale was fractured and/or fragmented from the explosive work. Three of these tests indicated either formation damage and/or increased fracturing of the shale existed between wells.

Table IV.  Shooting Data, Third Shot, Small Five-Spot
Pattern, Green River Site 1

| Well No. | Total Depth of Well (ft) | Total Depth of Casing (ft) | Explosive Depth (ft) | Sand Tamp Depth (ft) | Explosive Used (lb) |
|---|---|---|---|---|---|
| 2 | 381.0 | 342.0 | 346 | 190 | 780 |
| 3 | 384.0 | 343.0 | 345 | 200 | 1920 |
| 4 | 379.0 | 341.0 | 351 | 252 | 1800 |
| 5 | 383.0 | 341.0 | 346 | 197 | 1320 |
| 6 | 378.0 | 342.0 | 350[a] | 200 | 1320 |

[a] Stopped pouring explosive because of bridging with excessive water influx.

### Conclusions

Results of explosive fracturing tests in oil shale show that NGl will detonate and that the explosion will propagate in water-filled natural fractures and sand-propped, hydraulically induced fractures in oil shale. The shale was fragmented by this method, and a successful underground retorting experiment to recover shale oil was performed.

A combination of displacing NGl into a natural fracture system and using pelletized TNT in wellbore shots fragmented oil shale between wells at relatively shallow depths ranging from 60 to 100 ft. Extensive fragmentation extending to a radius of approximately 48 ft and extensive fractures to a radius of 90 ft were disclosed by various evaluation methods.

Further, pelletized TNT performed satisfactorily in wellbore shots in wells 150–385 ft deep. Fractures were created between wells as indicated by airflow tests, but numerous other evaluation techniques did not indicate the extent of rock fragmentation.

### Literature Cited

1. Nicklen, C. O., "History-Making Blast Set Off February 4," *Oil Bull*. (Feb. 8, 1946) **432**, 1.
2. Brewer, B., "Stimulation of Oil Production by the Use of Explosives After Hydraulic Fracturing," *Prod. Mon.* (February 1957) **21**, (4), 22.
3. Zandmer, S. M., "Method of Treating Oil and Gas Wells," U.S. Pat. **2,246,611** (Oct. 12, 1936).

4. "Pressure Reduction Chamber and Unloading Valve for Explosives," U.S. Pat. **2,504,611** (Feb. 25, 1946).
5. Brandon, C. W., "Method of Explosively Fracturing a Productive Oil and Gas Formation," U.S. Pat. **3,066,733** (Dec. 4, 1962).
6. Hanson, A. W., "Plastically Deformable Solids in Treating Subterranean Formations," U.S. Pat. **3,159,217** (Dec. 1, 1964).
7. Hinson, F. R., "Method and Apparatus for Treating an Earth Formation Penetrated by a Well," U.S. Pat. **3,191,678** (Apr. 2, 1962).
8. "Space-Age Explosive May Revive Well-Shooting," *Oil Gas J.* (Sept. 19, 1966) **64**, (38), 82.
9. Howell, W. D., Eakin, J. L., Miller, J. S., Walker, C. J., "Nitroglycerin Tests Prove New Application," *World Oil* (November 1970) **171** (6), 96.
10. Childs, O. E., "The Status of the Oil Shale Problem," *Colo. Sch. Mines Quart.* (July 1965) **60**, (3), 1.
11. Campbell, G. G., Scott, W. G., Miller, J. S., "Evaluation of Oil-Shale Fracturing Tests Near Rock Springs, Wyo.," *U.S. Bur. Mines Rep. Invest.* (1970) **7397**.
12. Miller, J. S., Nicholls, H. R., "Methods and Evaluation of Explosive Fracturing in Oil Shale," *U.S. Bur. Mines Rep. Invest.* (1973) **7729**.
13. Miller, J. S., Walker, C. J., Eakin, J. L., "Fracturing Oil Shale With Explosives for *In Situ* Oil Recovery," *U.S. Bur. Mines Rep. Invest.* (1974) **7874**.
14. Carpenter, H. C., Burwell, E. L., Sohns, H. W., "Evaluation of an *In Situ* Retorting Experiment in Green River Oil Shale," *J. Petrol. Technol.* (January 1972) **24**, (1), 21.
15. Grant, B. F., "Retorting Oil Shale Underground—Problems and Possibilities," *Colo. Sch. Mines Quart.* (July 1964) **59**, (3), 39.
16. Hill, G. R., Dugan, P., "The Characteristics of a Low Temperature *In Situ* Shale Oil," *Colo. Sch. Mines Quart.* (July 1967) **62**, (3), 75.
17. Miller, J. S., Howell, W. D., "Explosive Fracturing Tested in Oil Shale," *Colo. Sch. Mines Quart.* (July 1967) **62**, (3), 63.
18. Burwell, E. L., Carpenter, H. C., Sohns, H. W., "Experimental *In Situ* Retorting of Oil Shale at Rock Springs, Wyo.," *U.S. Bur. Mines Tech. Prog. Rep.* (June 1969) 16.
19. Miller, J. S., Howell, W. D., Eakin, J. L., Inman, E. R., "Factors Affecting Detonation Velocities of Desensitized Nitroglycerin in Simulated Underground Fractures," *U.S. Bur. Mines Rep. Invest.* (1969) **7277**.

RECEIVED December 16, 1974.

# 9

# Development of Communication Paths Within a Tar Sand Bed

D. A. REDFORD

Product Research and Development, Alberta Research Council, Edmonton, Alberta, Canada

P. F. COTSWORTH

Petrofina Canada Limited, Calgary, Alberta, Canada

*The* in situ *methods for recovering bitumen from the Athabasca tar sands depend upon successful development of a useable communication path. The major problem is blockage of the path caused by slumping of the tar sand or cooling of the hot unemulsified bitumen when it enters a cold region. Blockage is particularly acute in the early stages of path development. There are several in* situ *recovery methods, as well as two new approaches based on low-temperature emulsification. Certain agents promote emulsification at formation temperature (40°F). The other approach uses chemical reagents which react with the bitumen in the formation, forming water-soluble surfactants. Subsequent injection of water results in their ready displacement, thus enlarging the path and alleviating blockage.*

One of the first steps to understanding *in situ* recovery of bitumen from the Athabasca tar sands is fully to appreciate the fact that, under reservoir conditions, the bitumen in the tar sands cannot be made to flow under the influence of any reasonable or practical pressure gradient. This fact forces a change in the traditional concepts of secondary oil field recovery and necessitates the introduction of new approaches. These approaches often require that much of our traditional thinking regarding secondary oil recovery be changed and in some cases reversed.

If, because of its extremely high viscosity, the bitumen cannot be moved ahead of a front, then it must either be changed and then moved by a front, or it must be nibbled at from one side and the products carried away to a production point. The first approach has apparently been successfully applied by Pan American Petroleum Co. in their operations at Gregoire Lake (*1*). There, the tar-bearing McMurray formation is first heated to about 200°F, and then the preheated bitumen is driven to production wells by a combination of forward combustion and water flooding (COFCAW) process). The second approach has been demonstrated by laboratory and field work carried out by the Shell Oil Co. during the period 1956–1962 (*2–11*). This approach, referred to as the Shell process, involves three main steps:

1. Drilling to the base of the McMurray formation a series of production and injection wells in some suitable pattern. These wells are cased such that injection or production of fluids takes place only at or near the bottom of the tar–sand interval.

2. Achieving initial interwell communications between production and injection wells by horizontal hydraulic fracturing. The initial communications are then developed to the point of accepting large volumes of steam without sealing. This is done by emulsifying the bitumen along the fracture path through the action of critical concentrations of sodium hydroxide and heat, forming a low-viscosity oil-in-water (o/w) emulsion which is then transported to the production well.

3. Injection of large quantities of steam to the well developed communications path to achieve principal bitumen production. This steam moves up the formation, heating and emulsifying the cold bitumen above it, and is itself condensed. The condensate and emulsified bitumen form an oil-in-water emulsion which trickles down the formation and is driven to the production wells by pressure of the injected steam.

Of these three steps, the most difficult to achieve in the field is the enlargement of interwell communications paths, particularly the development of a cold fracture path into a hot communications path which will accept large quantities of steam without sealing. At formation temperatures ($\sim$ 40°F for 200–300 ft of overburden), the bitumen binds the formation sand together to form an almost brittle solid mass which fractures when subjected to a parting pressure. However, as the temperature increases, the formation softens and becomes an amorphous solid. Still further heating causes the bitumen to flow as a viscous heavy crude (6–7 API°). If steam is applied directly to a fracture path, propped or otherwise, the formation first softens and tends to slump into the fracture and may result in sealing. Further heating causes the unemulsified bitumen to flow. This unemulsified bitumen is moved further along the fracture path where it contacts colder areas of the formation and cools to form a highly viscous impermeable plug. This problem was recognized by Shell (*6*) during their experimental field work. Their solution (*6, 9, 10*) was to inject steam at a pressure above the pressure required to support the overburden (fracture propping pressure) but below the pressure which would propagate a vertical fracture. With this approach the injection pressure gradually rose, and from time to time it became necessary to replace the steam injection with injection of a hot solution of a critical concentration of sodium hydroxide. By repeating this cycle over a period of time, Shell was apparently able to develop hot communications paths which would accept large quantities of steam without sealing.

In pilot plant and field experiments in which we have been associated, this procedure has been less than satisfactory. Our experience has been that communications could not be maintained when using injection pressures below the vertical-fracture pressure. Use of pressures above the vertical-fracture pressure resulted in surface fractures and/or loss of fluids to high lying relatively permeable zones. This problem is particularly acute in areas where overburden is relatively light (*e.g.*, 200–300 ft). To overcome these difficulties, it was suggested that initial emulsification should take place at formation temperature (*i.e.*, cold emulsification) and that heating of the communications path take place at a rate such that all the bitumen entering the path would be emulsified. By this method the propped communications path would never become plugged with unemulsified bitumen, and a low pressure process could be developed for obtaining a hot communications path through a tar sand bed.

### Results and Discussion

It has been our experience and that of Shell (8) that sodium hydroxide and other bases are ineffective in promoting the formation of oil-in-water emulsions below 60°F. Their ability to promote emulsion formation gradually increases with temperature but does not become really effective until 90°F or more. Athabasca tar sands, however, show considerable softening at 60°F and will begin to weep bitumen at 90°F. It was, therefore, important to find emulsifying agents which were more effective at the lower temperatures. It was found that a combination of sodium hydroxide and a nonionic surfactant (T–45, manufactured by Rohm and Haas Co., an octylphenoxy polyethylene oxyethanol in which the octyl group is branched and which contains about 5 moles of ethylene oxide) was effective in promoting low temperature emulsification of the bitumen. Optimum concentrations of this surfactant and sodium hydroxide as a function of temperature as determined from our experimental work are given in Table I. Using these concentrations and a gradual heating of the injected fluids, good hot communications were developed over a 100-ft interval near the base of the McMurray formation in a period of about six weeks. Pressures were maintained below the fracture-

**Table I.   Optimum Concentrations of TX45 and NaOH as a Function of Temperature**

| Temperature (°F) | TX45 Concentration (%) | NaOH Concentration (%) |
|---|---|---|
| 40– 60 | 0.4 | 0.2 |
| 60– 70 | 0.2 | 0.2 |
| 70– 80 | 0.2 | 0.15 |
| 80–100 | 0.1 | 0.15 |
| 100–120 | 0.1 | 0.1 |
| 120–200 | — | 0.1 |
| > 200 | — | — |

propping pressure. Stable emulsions were achieved, ranging in concentration from 3% for the cold emulsification up to 17% for the warmer emulsification.

Despite these encouraging results, it was of interest to find other agents which would produce more concentrated cold-temperature emulsions. It was believed that the sodium hydroxide reacted with the acidic functional groups in the bitumen to form organic salts and that these organic salts then acted as surface-active agents which caused emulsification of the bitumen. It was, therefore, hoped that the amount of these surface-active agents could be increased by oxidizing part of the bitumen to form more acidic groups. Several oxidizing agents were tried; ozone was the most successful.

### Experimental

Ozone (6% in oxygen) was passed for two days through a loosely packed vertical column of Athabasca tar sand and the column eluted

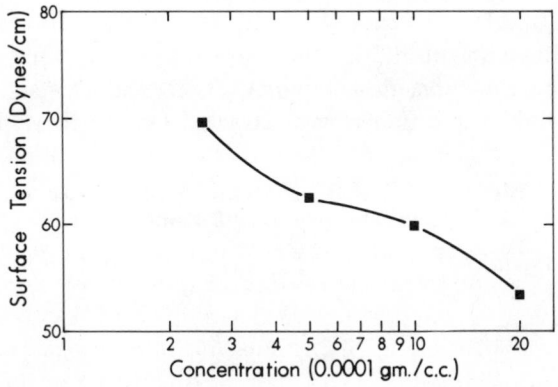

*Figure 1.   Surface tension of water soluble fraction of ozonized bitumen*

**Table II.    Analysis of Unozonized and Partly-Ozonized Bitumen**

| Sample | Carbon (%) | Hydrogen (%) | Oxygen (%) | Sulfur (%) | Nitrogen (%) | Ash (%) | Molecular Weight |
|--------|-----------|--------------|------------|-----------|--------------|---------|------------------|
| O | 81.82 | 10.37 | 0.78 | 5.17 | 1.23 | — | 1448 |
| A | 81.63 | 9.98 | 2.18 | 4.02 | 1.28 | 1.11 | 668 |
| B | 81.71 | 10.62 | 2.25 | 3.97 | 0.88 | 0.92 | 625 |
| C | 81.66 | 10.67 | 2.37 | 3.95 | 0.71 | — | 634 |
| D | 81.44 | 10.55 | 2.25 | 4.30 | 1.05 | 1.05 | 641 |

with water. The tar sand changed markedly in appearance, revealing many white sand grains while the eluted material was a dark brown color and foamed when lightly shaken. Evaporation of the eluent to dryness yielded a material which analyzed: carbon, 45.3%; hydrogen, 5.5%; oxygen, 40.6%; nitrogen, 1.4%; and sulfur, 7.2%. When stirred with water, under a microscope, the sand of the ozonized tar sands separated to form white water-wet grains while the bitumen formed globules in the water phase. Similarly treated unozonized sands produced no noticeable change.

A 3-ft by 2-in. horizontal tube was then tightly packed with bitumen, and a narrow path of 20/40-mesh Ottawa sand was placed at the bottom. Ozone (6% in oxygen) was passed through the tube for two and a half days. The exit gas contained ~ 1% ozone. A 50-g sample of the tar sand was extracted with water (500 ml) to yield a dark brown solution which foamed when lightly shaken. A surface-tension curve vs. concentration for this sample is given in Figure 1. Evaporation of the solution to dryness yielded 0.237 g of material containing: carbon, 28.7%; hydrogen, 3.7%; oxygen, 51.3%; nitrogen, 1.3%; and sulfur, 8.2%. A second similar 50-g sample of tar sand was extracted with water, and the resulting solution was neutralized to pH 7 with 0.1N sodium hydroxide (41.4 ml required).

In sampling the tube, samples were taken from four equal length sections along the tube, designated A, B, C, and D starting from the ozone inlet end. The bitumen was extracted from each of these samples

**Table III.    S.A.R.A. Analysis of Unozonized and Partly-Ozonized Bitumen**

| Sample | Asphaltenes (%) | Resins I (%) | Resins II (%) | Saturates (%) | Aromatics (%) |
|--------|-----------------|--------------|---------------|---------------|---------------|
| O | 21.8 | 42.3 | 4.2 | 15.9 | 14.3 |
| A | 24.7 | 43.9 | 4.3 | 13.3 | 10.6 |
| B | 25.3 | 42.1 | 4.0 | 18.4 | 8.8 |
| C | 25.8 | 39.5 | 2.5 | 20.0 | 9.0 |
| D | 24.5 | 44.8 | 2.6 | 18.7 | 6.7 |

and from a sample of unozonized tar sand (designated sample O) using toluene reflux. Elemental analyses of these bitumen samples and their molecular weights are in Table II. Standard S.A.R.A. (saturates, aromatics, resins, and asphaltenes) analyses were conducted on each of the

**Table IV.   Analysis of Unozonized and Partly-Ozonized Bitumens After S.A.R.A. Analysis**

| Sample | Carbon (%) | Hydrogen (%) | Oxygen (%) | Sulfur (%) | Nitrogen (%) | Ash (%) | Molecular Weight |
|--------|--------|--------|--------|--------|--------|--------|--------|
| Asphaltenes | | | | | | | |
| O | 78.84 | 7.83 | 3.03 | 8.48 | 1.34 | 0.64 | 4797 |
| A | 77.60 | 7.90 | 4.11 | 7.88 | 1.81 | 0.79 | 4722 |
| B | 76.86 | 7.74 | 4.88 | 7.68 | 0.82 | 1.88 | — |
| C | 76.87 | 7.65 | 4.90 | 7.75 | 0.90 | 1.80 | — |
| D | 77.38 | 7.88 | 4.44 | 7.71 | 0.89 | 1.53 | 3493 |
| Resins I | | | | | | | |
| O | 76.89 | 9.72 | 3.97 | 5.48 | 0.73 | 1.01 | 727 |
| A | 77.27 | 9.48 | 4.23 | 4.77 | 0.77 | 1.31 | 731 |
| B | 78.53 | 9.73 | 6.05 | 4.65 | 0.25 | — | 563 |
| C | 78.71 | 9.63 | 6.05 | 4.71 | 0.26 | — | 590 |
| D | 78.67 | 9.73 | 5.97 | 4.64 | 0.17 | — | 617 |
| Resins II | | | | | | | |
| O | 76.31 | 9.44 | 4.44 | 2.30 | 0.30 | 2.17 | — |
| A | 77.47 | 9.89 | 6.03 | 1.17 | 1.17 | 1.15 | — |
| B | 78.99 | 10.50 | 7.84 | 1.96 | 0.18 | — | — |
| C | 79.37 | 9.53 | 9.07 | 1.62 | 0.21 | — | — |
| D | 79.17 | 10.63 | 8.00 | 1.71 | 0.19 | — | — |
| Saturates | | | | | | | |
| O | 85.81 | 13.31 | 0.17 | 0.28 | 0.45 | — | 454 |
| A | 85.92 | 13.34 | 0.14 | 0.27 | 0.41 | — | 433 |
| B | 86.31 | 13.03 | 0.08 | 0.26 | 0.18 | — | 397 |
| C | 86.57 | 12.88 | 0.22 | 0.36 | 0.21 | — | 395 |
| D | 86.28 | 13.10 | 0.15 | 0.25 | 0.13 | — | 412 |
| Aromatics | | | | | | | |
| O | 85.13 | 10.33 | 0.27 | 3.63 | 0.59 | — | 398 |
| A | 84.94 | 10.25 | 0.19 | 3.75 | 0.64 | — | 397 |
| B | 85.33 | 10.61 | 0.45 | 3.13 | 0.15 | — | 390 |
| C | 85.30 | 10.45 | 0.48 | — | 0.16 | — | 373 |
| D | 85.29 | 10.48 | 0.50 | 3.22 | 0.26 | — | 379 |

samples; thus dividing each of them into asphaltenes, resin I, resin II, saturates, and aromatics. Results of this analysis are given in Table III. Elemental analyses and molecular weights were done on each of the subsamples, and the results are shown in Table IV. These analyses show that in general ozonolysis occurred to the asphaltene and resin fractions of the bitumen and resulted in a lower molecular weight and a decreased sulfur content. The aromatic fraction was also affected. However, since it comprises only about 10% of the bitumen, this had little affect on the overall molecular weight or viscosity. The saturates which are low in molecular weight and viscosity were only slightly affected. Thus, ozonolysis in attacking the aromatic double bonds is mainly attacking that part of the bitumen which is of highest molecular weight and viscosity and converting it into water-soluble or more hydrophylic material.

A second 3-ft by 2-in. horizontal tube was packed, (and ozone (6% in oxygen) was passed through for seven days (80% of the ozone was still being absorbed by the cell after seven days). Distilled water (20 ml/hr) and ozone (6% in oxygen) were then passed through the tube for six days. Examination of the cell disclosed areas around the sand

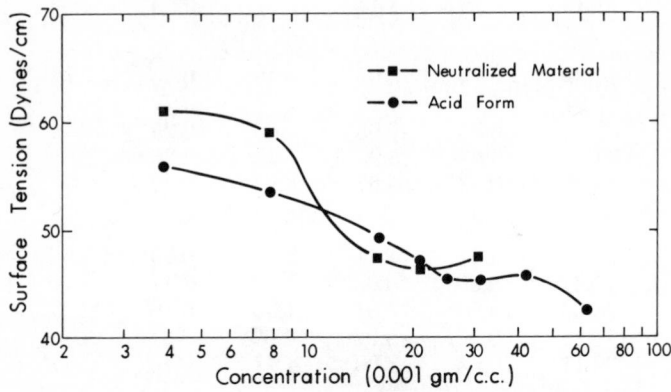

*Figure 2.    Surface tension of water soluble ozonized bitumen*

path and extending out around the surface of the cell which had been largely depleted of bitumen, leaving clean white sand. The initial effluent from the cell (first day's effluent after commencing water injection) was an amber color (pH 1.35) and contained about 6% water-soluble organic material. This material exhibited a surface-tension curve as given in Figure 2. The surface tension of the neutralized material as a function of concentration is also given in Figure 2. Approximately 9.5% of the bitumen originally in the cell was removed during the six days of water injection.

The test cell was reassembled, and a solution of 0.2% sodium hydroxide was passed through the cell for 14 days (20 ml/hr). An additional 8.7% of the bitumen in the cell was removed during this period. Most of the cell effluent which was collected during the first six days of water injection was neutralized to pH 7, and this material was recycled through the cell six times. This procedure removed a further 2% of the bitumen from the cell, for a total bitumen recovery of 20.2%. When the cell was opened and the tar sand examined, it became evident that the whole cross section of the tube had been affected to some extent by the combination of ozonolysis, distilled water flush, sodium hydroxide solution flush, and neutralized effluent recycle, but in some areas the bitumen had been extensively removed, especially near the original sand path and extending around the glass surface.

These results indicate that ozone readily reacts with bitumen at formation temperatures ($40°F$) to form water-soluble highly oxygenated materials which have surface-active properties in both acid form and as neutralized salts. Passing water through an ozonized formation of tar sands results in removal of part of the bitumen; in some areas this removal is extensive while in other areas it is minor. When 0.2% sodium hydroxide solution is passed through such a bed, more bitumen is removed and the highly depleted areas are extended. It is our intention to test a pilot-plant simulation of field conditions for this process to determine whether the combination of ozonolysis, water flush, 0.2% sodium hydroxide flush, and gradual heating can be used for the low pressure development of a cold propped fracture into a hot communications path which will accept steam without sealing.

## Conclusion

It has been demonstrated in the field and in the laboratory that by using a combination of nonionic surfactant and base (sodium hydroxide) and using pressures substantially below the propping pressure, a cold propped fracture path can be developed into a zone that will accept large volumes of steam without sealing. Further, in the laboratory, that ozone will readily react with bitumen at formation temperatures, to form highly oxygenated water-soluble surface-active agents. Therefore, it is expected that ozone together with water and/or dilute basic solutions can be used at low pressures to expand a propped fracture path into a hot communications path which will accept steam without sealing.

## Literature Cited

1. Muskeg Oil Co., Application No. 4301 to the Oil and Gas Conservation Board of the Province of Alberta, Canada, October 1968; hearings Jan. 22, 1969.
2. Shell Oil Co. of Canada Ltd., Application to the Oil and Gas Conservation Board of the Province of Alberta for the Approval of a Scheme or Operation for the Recovery of Oil or a Crude Hydrocarbon Product from the Athabasca Oil Sands, Feb., 1963.

3. Doscher, T. M., "Technical Problems in *In Situ* Methods for Recovery of Bitumen from the Tar Sands," Panel Discussion No. 13, 7th World Petroleum Congress.
4. Doscher, T. M., Labelle, R. W., Sawatsky, L. H., Zwicky, R. W., "Steam Drive—A Process for *In Situ* Recovery of Oil from the Athabasca Oil Sands," Athabasca Oil Sands K. A. Clark Volume, *Res. Coun. Alberta Inf. Ser.* 45 (Oct., 1963).
5. Doscher, T. M., Oil Recovery, Canadian patent 711,556 (June 15, 1965).
6. Closmann, P. J., Doscher, T. M., Matthews, C. S., Recovery of Viscous Petroleum Materials, U.S. Patent 3,221,813 (Dec. 7, 1965).
7. Doscher, T. M. *et al.*, Oil Recovery from Tar Sands, U.S. patent 2,882,973 (Apr. 21, 1959).
8. Doscher, T. M. *et al.*, Oil Recovery from Tar Sands, Canadian patent 639,050 (March 27, 1962).
9. Matthews, C. S. *et al.*, Thermally Controlling Fracturing, U.S. patent 3,379,250 (Apr. 23, 1968).
10. Doscher, T. M., Oil Recovery, U.S. patent 3,279,538 (Oct. 18, 1966).
11. van Meurs, P., Volek, C. W., Steam Drive for Incompetent Tar Sands, U.S. patent 3,396,791 (Aug. 13, 1966).
12. Redford, D. A., Process for Developing Interwell Communications in a Tar Sand, U.S. patent 3,706,341 (Dec. 19, 1972); Canadian patent 933,343, related.

RECEIVED December 16, 1974.

# Characterization of a Utah Tar Sand Bitumen

J. W. BUNGER

Laramie Energy Research Center, Bureau of Mines, U.S. Department of the Interior, Laramie, Wyo. 82070

*A preliminary characterization of a Utah tar sand bitumen has been made using methods developed for high boiling petroleum fractions. The characterization includes information about the major compound types which can be compared with similar data for other tar sand bitumens and, more importantly, can be correlated with data from petroleum samples for which refining characteristics are known. Examination of the P. R. Spring bitumen showed that it differed significantly from representative petroleum residues, principally in its high nonhydrocarbon content. Compositional information is important because of the effects that composition has on the recovery and processing of the bitumen.*

Until recently, tar sand deposits have been relatively unimportant as an energy resource. However, increasing demands for energy have prompted a greater effort to use the energy stored in bituminous sandstone deposits. The inaccessibility of the bitumen, which is generally impregnated in subsurface sandstone, and the high viscosity of the bitumen cause major recovery problems. The high viscosity, heteroatom content, and molecular weight will probably cause problems with utilization. Knowledge of the composition of tar sand bitumens would facilitate accurate prediction of the chemical and physical behavior of these bitumens in recovery and refining processes.

Recently, studies have been made to determine the properties of Utah bitumens (1, 2). Others (3-9) have studied the properties of Athabasca bitumens. The analyses of the bitumens have generally included physical properties, elemental analyses, distillation, and infrared and nuclear magnetic resonance spectroscopy. These analyses have been carried out on total bitumens and subfractions generally defined by solubility and/or adsorption chromatography. Detailed analysis of the chemical composition of a total bitumen and of subfractions defined in this manner is difficult because of the complexity of the samples. A preliminary separation defined by chemical functionality would provide simplified fractions for further separation and analysis.

This study presents the results of a preliminary examination for compound types present in a P. R. Spring, Utah, tar sand bitumen. Physical properties, elemental analysis, and distillation data are given and compared with literature values for other tar sand bitumens. The separation of the bitumen, using selected techniques developed in our laboratory (10, 11, 12, 13, 14), into acid, base, neutral Lewis bases (formerly called neutral nitrogen), saturated, and aromatic hydrocarbon fractions is described. The analytical results of the separation are compared with those for high boiling petroleum residues separated in a similar fashion. This compositional information could provide an evaluation of the bitumen for processing because more is known of the processing characteristics of petroleum fractions than of tar sand bitumens.

## Experimental

**Description of Bitumen Sample.** The core that was extracted to produce the bitumen came from Colvert No. 1, NW1/4SE1/4, Sec. 35, T. 15S., R. 22 E., Uintah County, Utah. The tar-bearing sand occurred in the P. R. Spring deposit, Douglas Creek Member, Green River Formation of Tertiary age. Core samples from Colvert No. 1 showed that tar-bearing sand occurred intermittently between the depths of 64 and 162 ft. Approximately 2-in. sections were taken from each foot between 84 and 117 ft, combined, and exhaustively extracted with benzene in a Soxhlet extractor. The benzene extract was filtered through a 4.0–5.5 $\mu$ fritted glass disk funnel, and the benzene was removed by rotary evaporation (75–80°C, 4–5 torr). The recovered bitumen was used for property measurements and as the starting material for the separation into defined fractions.

**Simulated Distillation of Bitumen Sample.** The boiling-range distribution of the recovered bitumen was determined by simulated distillation gas–liquid chromatography using the procedure of Poulson et al. (15). Boiling points are determined by calibration with a mixture of $n$-paraffins ranging from $C_{11}$ to $C_{42}$. The upper limit for boiling point determination in this analysis is about 540°C (1000°F).

**Separation of the Bitumen into Defined Fractions.** The procedure was an extension of one developed for the separation of high boiling petroleum cuts into five fractions—acids, bases, neutral Lewis bases, saturates, and aromatics (*10*). This separation and the further division of the first three fractions into subfractions is shown schematically in Figure 1.

A 20-g sample of the bitumen was dissolved in cyclohexane and charged to a column containing 50 g of Amberlyst A-29 anion exchange resin (OH⁻ form) on top of 50 g of Amberlyst 15 cation exchange resin (H⁺ form) (Rohm and Haas). The column was exhaustively eluted with

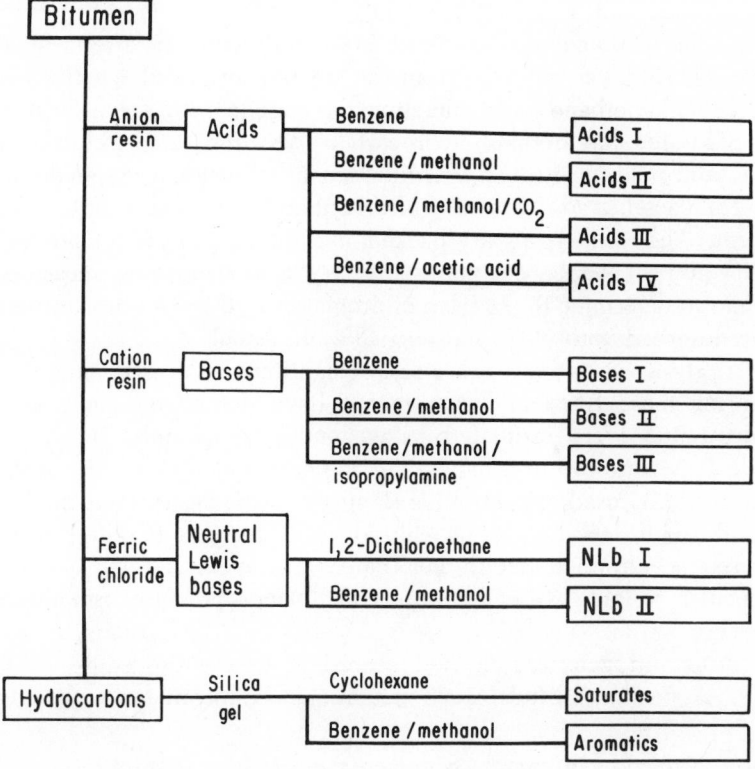

*Figure 1. Separation and further division of first three fractions of the bitumen*

cyclohexane to remove nonreactive material. The resins were removed from the column, separated, and placed in individual water-jacketed Soxhlet extractors. The acids were extracted sequentially from the anion resin with benzene, 60% benzene–40% methanol, 60% benzene–40% methanol saturated with carbon dioxide at 0°C, and 5% acetic acid–95% benzene. These last two solvent mixtures were applied manually rather than by the usual Soxhlet reflux because they do not form azeotropes. The bases were removed from the cation resin in a similar fashion using the solvents benzene, 60% benzene–40% methanol, and 8% isopropyl amine–55% benzene–37% methanol. All percentages are on a volume basis. These seven subfractions were retained separately for analysis and labeled, respectively, acids I, II, III, and IV and bases I, II, and III.

The neutral Lewis bases were removed from the acid- and base-free bitumen by chromatography in cyclohexane on ferric chloride/Attapulgus clay, packed in a column above Amberlyst A-29. The ratio of sample to ferric chloride/Attapulgus clay to anion resin was 1:13:17. Neutral Lewis bases were recovered by successive column elution with 1,2-dichloroethane and 40% methanol–60% benzene to provide two subfractions. The methanol was removed from subfraction II, after which the ferric chloride was removed from the fraction by dissolving the sample in 1,2-dichloroethane and contacting the sample with the A-29 resin.

Saturated and aromatic hydrocarbons were separated from the acid-, base-, and neutral nitrogen-free bitumen by adsorption chromatography using silica gel, grade 12, as the adsorbent and cyclohexane as the eluting solvent. The column was dry packed, and the cutpoint was made at two void volumes. At the cut points, the UV absorbance was measured at 270 nm to determine the overlap of aromatics in the saturates. Aromatics were desorbed with 60% benzene–40% methanol.

**Analysis of Defined Fractions.** Quantitative infrared analysis was used for those fractional groups that have definitive bands; average absorptivities were estimated using model compounds (12, 16, 17). Table I lists the infrared bands and the apparent integrated absorption intensities (B) used. Quantitative IR spectra were measured in methylene chloride with 0.05 cm sodium chloride cells, using a Perkin–Elmer 521 infrared spectrophotometer. Peak area was measured by planimetry. Molecular weights were determined by vapor–pressure osmometry in benzene.

Basic nitrogen was titrated according to the procedure of Buell (18). A Beckman Model 1063 potentiometric titrimeter with calomel

**Table I. Infrared Assignments and Apparent Integrated Absorption Intensity**

|  | Wavenumber (cm$^{-1}$) | B(l./mol-cm$^2 \times 10^4$) |
|---|---|---|
| Phenols | 3540–3600 | 0.5[a] |
| Carbazoles | 3455–3465 | 0.7 |
| Carboxylic acids | 1700–1745 | 1.5[a] |
| Ketones | 1690–1700 | 0.7[b] |
| Amides | 1625–1690 | 1.5 |
| Sulfoxides | 1025–1040 | 0.6[c] |

[a] Ref. 12.
[b] Ref. 17.
[c] Ref. 16.

Table II.    Properties of P.R. Spring Bitumen

| | |
|---|---:|
| Specific gravity 60/60 | .998 |
| API gravity | 10.3 |
| Penetration (50 g) | 130 |
| Viscosity 77°F (poise) | 325,000 |
| Heating value (Btu/lb) | 18,000 |
| Carbon (%) | 84.44 |
| Hydrogen (%) | 11.05 |
| Sulfur (%) | 0.75 |
| Nitrogen (%) | 1.00 |
| Ash (%) | 0.17 |
| C/H (atomic ratio) | .637 |

and glass electrodes was used to titrate the bitumen solution with 70% perchloric acid in dioxane. The solvent system was 2:1 benzene:acetic anhydride. Endpoints were determined at the inflection point of the curve. Calibration data allowed calculation of the percentage of titratable nitrogen in the sample. The half-neutralization potential (HNP), the potential of the curve midway through the titration, provides information on the strength of the bases.

Saturated hydrocarbons were examined with a CEC-21-110B mass spectrometer. The standard ASTM method (*19*) was used to classify compounds according to structure.

### Results and Discussion

**Properties of the Bitumen.** Specific and API gravities, viscosity, heating value, and elemental analysis for this bitumen are given in Table II. The sandstone core from which the bitumen was extracted remained consolidated throughout the extraction. The extracted organics represented 5.27 wt % of the core. The filtration of the bitumen to remove fine inorganic materials was adequate, as evidenced by ash content of only 0.17 wt %. The gravity (10.3 API°) is average for Utah tar sand bitumens and is slightly higher than for Athabasca bitumens which average 7.6 API° (Ref. 3, p. 23). The viscosity of the study bitumen was estimated to be 325,000 poise by cone–plate viscometry at a temperature of 77°F and a shear rate of 0.05 sec$^{-1}$. Conventional viscosity measurements were not applicable because the bitumen was semi-solid at room temperature and only slightly fluid at 80°C. Standardized methods for the bitumen workup and viscosity determination have not been developed. Such methods must recognize the effect that entrained solvents and the loss of light ends during solvent removal have on viscosity. Camp (*3*) indicates that variations in these two phenomena are significant factors in the variations of viscosities from sample to sample. The heating value of 18,100 Btu/lb is about average for other P. R. Spring samples measured in our laboratories and slightly higher than for Athabasca bitumens which average 17,800 Btu/lb. A sulfur content of 0.75% is slightly higher than for other bitumens of the Uinta Basin but considerably lower than that for the Tar Sand Triangle, Utah, and the Athabasca, Canada, bitumens which typically contain 2–6% sulfur. A nitrogen content of 1.00 is average for Uinta Basin tars, but is higher than for other Utah and Athabasca bitumens which typically contain 0.35–0.65%. The carbon–hydrogen ratio of 0.637 is average for Utah bitumens and slightly lower than for Athabasca bitumens which average 0.699.

The property data show that the study bitumen is typical for a Uinta Basin tar sand bitumen; Uinta Basin bitumens are atypically low in sulfur content compared with other Utah and Athabasca bitumens. Utah bitumens are slightly higher in API gravity, in heating value, and in hydrogen content than are Athabasca bitumens. The slightly higher hydrogen content of the Utah bitumens tends to increase their relative value over the Athabasca bitumens as a hydrocarbon fuel source.

**Boiling Point Distribution—Simulated Distillation.** The boiling point distribution of the material in the bitumen was determined using simulated distillation by gas–liquid chromatography. This technique has several advantages over conventional assay distillations. It is rapid and can analyze very small samples (30 mg). The boiling point determination is extended by 120°C above that of the standard Bureau of Mines crude oil analysis (BMCOA), giving more information about the bitumen. The relatively short contact time at high temperatures minimizes the possibility of thermal cracking, which produces artifacts and alters the distillation curve. The amount of solvent remaining in the bitumen can be determined accurately so that its effects on other properties can be calculated or estimated. Complete removal of the extracting solvent is nearly impossible, and its presence in the bitumen affects the amount of recovery, the elemental analysis, and, especially, the viscosity.

Simulated distillation affords the option of conveniently choosing cutpoints for calculations of the boiling range distribution. For purposes of comparison with other bitumens, cutpoints were chosen to coincide with the fraction for the BMCOA distillations. The gas–liquid chromatography conditions used for the present analysis are designed to analyze the high boiling portions and will not accurately resolve constituents boiling below 125°C. Hence, fractions 1–4 were combined in the calculation.

The simulated distillation results are given in Table III, column A, along with literature values for the distillations of five other P. R. Spring samples. The 0.4% material appearing in fraction 1–4 for this study

bitumen is benzene left from the extraction procedure. For the first 15
fractions, the simulated distillation results approximate those reported
by Gwynn (2) in that no material boils below 250°C and the percentage
of > 420°C residue is high. The values reported by Wood and Ritzma
(1), however, vary significantly in that substantial amounts of light ends
are present with very little material boiling between 275° and 420°C.
These values suggest that thermal cracking occured during the atmospheric distillation stage; thus, the large fractions below 275°C are prob-

**Table III.   Distillation Data for P.R. Spring Bitumens**

*Wt %*

| Fraction | Cut Point (°C) | This Study[a] (core) | Main Canyon Seep (2) | Core 79–83 Ft (2) | Core 137–141 Ft (2) | Outcrop Sample No. 69–13E (1) | Outcrop Sample No. 67–1A (1) |
|---|---|---|---|---|---|---|---|
| 1–4 | 125 | 0.4 | | | | 2.2[b] | |
| 5 | 150 | | | | | 2.4 | |
| 6 | 175 | | | | | 1.2 | |
| 7 | 200 | | | | | 1.2 | 1.4 |
| 8 | 225 | | | | | 1.4 | 6.5 |
| 9 | 250 | 0.5 | | | | 5.9 | 64.4 |
| 10 | 275 | 1.0 | | | | 31.1 | 3.1 |
| 11 | 308 | 2.1 | 1.9 | 1.6 | 3.1 | | |
| 12 | 336 | 2.8 | 2.3 | 2.2 | 2.8 | 0.2 | |
| 13 | 364 | 4.1 | 3.1 | 2.9 | 4.5 | 2.2 | |
| 14 | 392 | 3.2 | 3.4 | 5.6 | 4.1 | 2.5 | |
| 15 | 420 | 5.1 | 9.5 | 11.8 | 12.5 | 2.6 | 3.4 |
| Residue | | (80.8) | 75.8 | 74.8 | 71.0 | 47.1 | 21.3 |
| 16 | 448 | 6.4 | | | | | |
| 17 | 476 | 8.1 | | | | | |
| 18 | 504 | 7.1 | | | | | |
| 19 | 532 | 8.1 | | | | | |
| Residue | | 51.1 | | | | | |

[a] Simulated distillation data. All others are actual distillation.
[b] The total value of 2.2 was found in fraction 4.

ably cracked products that do not reflect the original composition of the
bitumen. These results point out the dangers and difficulties of obtaining
boiling point distributions of heavy bitumens by distillation and suggest
that a different approach such as simulated distillation should be used.

**Separation of the Bitumen.** The P. R. Spring bitumen was separated
according to the flow diagram in Figure 1. Acidic compounds were
isolated using a quaternary ammonium hydroxide anion exchange resin.
Because the system is nonaqueous, ion exchange does not occur; rather

Table IV.    Results of the Separation of Residue Samples

Residue Sample (Wt %)

| Fraction | P.R. Spring 225°C | Wilmington 485°C | Red Wash 545°C | Recluse 750°C | Gach Saran 675°C | Prudhoe Bay 675°C |
|---|---|---|---|---|---|---|
| Acids | 15.4 | 10.7 | 6.0 | 5.9 | 12.1 | 10.0 |
| Bases | 12.3 | 13.3 | 10.2 | 8.3 | 14.2 | 15.7 |
| Neutral Lewis bases | 18.5 | 20.4 | 10.8 | 17.4 | 23.2 | 12.6 |
| Saturated hydrocarbons | 25.7 | 18.4 | 51.8 | 40.8 | 25.6 | 32.9 |
| Aromatic hydrocarbons | 24.9 | 35.1 | 10.9 | 24.8 | 18.7 | 23.4 |
| Recovery | 96.8 | 97.9 | 89.7 | 97.2 | 93.8 | 94.6 |

an association between acidic types and the basic resin occurs. Similarly, the bases were separated because they associate with the sulfonic acid cation resin. The acids were removed from the resin sequentially by exhaustive elution with a series of solvents with increasing polarity; the strongest acids require the most polar desorbing solvent and would appear in the later fractions. The bases were removed from the cation resin with similar sequential elutions with solvents of increasing polarity. Thus, subfractions III and IV of the acids and subfraction III of the bases would contain the strongest acids and bases, respectively.

Neutral Lewis bases were removed by contacting the acid- and base-free bitumen with ferric chloride/Attapulgus clay in a column system. Weakly adsorbed complexes were desorbed with 1,2-dichloroethane while strongly held complexes were desorbed with benzene and methanol. This procedure provided two neutral Lewis bases fractions for analysis.

The name of the neutral nitrogen concentrate was changed to neutral Lewis bases in this publication for two reasons. First, new evidence indicates that not only nitrogen compounds, but also other electron donors such as ketones, sulfoxides, and esters adsorb or complex on the ferric chloride/Attapulgus clay system. These types are not generally present in petroleum distillates for which the separation was designed and the name "neutral nitrogen" derived. Hence, the term neutral nitrogen is too restrictive of the types which may be found in this fraction. Second, as the application of these techniques expands to shale oil, coal liquids, and synthetic crude oils, other electron donor types which are retained by this system are likely to be encountered. The term neutral Lewis bases encompasses all the types which might reasonably be found in this fraction.

The hydrocarbons portion, which contains some oxygen and sulfur, was separated into saturates and aromatics by adsorption chromatography on silica gel. Elution of the saturates with two void volumes of cyclohexane gave a satisfactory separation of saturates and aromatics. Previous experiments with this system have demonstrated that more than 99% of the saturates are eluted in two bed volumes at this cutpoint. The UV absorbance at 270 nm for the eluant in 1-cm cells was 0.20, indicating minimal overlap of aromatics in the saturates.

The initial separation, which provides the percentage of total acids, bases, neutral Lewis bases, saturates, and aromatics, gives information that could be useful in determining the value of a bitumen as a refining feedstock. Table IV lists data for these five major fractions from the P. R. Spring bitumen and compares it with similar data from five petroleum residues. Data on the petroleum residues were obtained by the procedure described (10) as an extension of our studies of heavy distillate fractions. Comparisons of the bitumen with these residues must be made with the recognition that the residues have higher initial boiling points than the bitumen; i.e., the bitumen contains more low boiling material. Examination of high boiling distillate cuts (10, 11, 13, 14) has shown that nonhydrocarbons (acids, bases, and neutral Lewis bases) are concentrated in the high boiling fractions. In general, the nonhydrocarbon content of the residue is twice that of the 500°–600°C cut, which is in turn twice that of the 400°–500°C cut. Table IV shows that the nonhydrocarbon content of the bitumen is high—second only to the Gach Saran residue—even though the bitumen contains 50% material in the 250°–500°C boiling range which is absent in the residues. This suggests that a comparable initial boiling point residue of the P. R. Spring bitumen would have an unusually high content of nonhydrocarbons.

The approximate nonhydrocarbon content can be calculated from the analytical results of the separation in conjunction with the distillation data. Table V lists the calculated nonhydrocarbon compound content for an average crude oil, a Wilmington crude oil, and the P. R. Spring

Table V.  Comparison of Nonhydrocarbon Content of
Petroleum and Tar Sand Samples

|  | 250°–535°C Distillate Nonhydrocarbon (Wt %) | > 535°C Residue Nonhydrocarbon (Wt %) |
|---|---|---|
| Typical crude oil[a] | 4 | 35 |
| Wilmington crude oil | 13 | 55 |
| P.R. Spring bitumen | 12 | 84 |

[a] Average for four crude oils other than Wilmington from Table IV.

bitumen. These values were calculated for the 250°–535°C distillate and for the residue. The calculation for the P. R. Spring sample was made as follows. The total hydrocarbons, which represent 50.6% of the bitumen, were 85% distillable ( < 535°C), thus 43% of the bitumen was distillable hydrocarbons. The total bitumen was 48.9% distillable, leaving 5.9% as distillable nonhydrocarbons. Therefore, the nonhydrocarbons represent 12% of the distillable portion. The nonhydrocarbon content of the nondistillable portion was calculated similarly. In the 250°–535°C distillate fraction, the nonhydrocarbon content of the P. R. Spring bitumen approximates that for the Wilmington crude oil which is atypically high in nonhydrocarbons. In the > 535°C residue, the P. R. Spring sample is considerably higher in nonhydrocarbons than the Wilmington sample. However, the total sulfur and nitrogen content of a > 535°C P. R. Spring residue is similar to that of the Wilmington > 485°C, indicating that the P. R. Spring heteroatomic compounds are of a higher average molecular weight. The unusually high nonhydrocarbon content of the P. R. Spring bitumen suggests that problems would be incurred in refining processes that are typically sensitive to nonhydrocarbons.

Another major difference between the tar sand bitumen and the petroleum residues is suggested in Table IV. In all the petroleum samples, the base content is higher than the acid content. In the P. R. Spring sample, the acids are higher than the bases. This could indicate the differences in oxidation, maturation, or origin for the tar sand bitumen as compared with crude oils. In addition, the acid content may often have important effects on recovery processes, such as those which rely on caustic flooding.

**Table VI.    Infrared Compound Type Analysis**

| Fraction | Wt % of Bitumen | Molecular Wt |
|---|---|---|
| Acids I | 4.8 | 1240 |
| Acids II | 2.2 | 1160 |
| Acids III | 4.0 | 850 |
| Acids IV | 4.4 | 850 |
| Bases I | 2.0 | 850 |
| Bases II | 1.5 | 1050 |
| Bases III | 8.8 | 950 |
| Neutral Lewis bases I | 10.3 | 980 |
| Neutral Lewis bases II | 8.2 | 2350 |
| Total (wt % of bitumen) | 46.2 | |

**Spectroscopic Analysis of Defined Fractions.** The eleven subfractions outlined in Figure 1 were examined by infrared spectroscopy. Only those functional groups that were characteristic and that were sufficiently resolved from other bands to be integrated were used (*see* Table I). To obtain quantitative data, molecular weights of the subfractions were measured by vapor pressure osmometry (VPO) in benzene. The accuracy of the molecular weights is subject to the degree of intermolecular association exhibited in the fractions. For very polar molecules such as strong acids or bases, this association could be considerable. For the carboxylic acid concentrates of petroleum samples, molecular weights determined by VPO have been about twice those determined by mass spectroscopy. Accurate determination of molecular weight of a concentrate from a total bitumen by mass spectroscopy is extremely difficult, however, because of the wide molecular weight (volatility) range represented. Along with the estimation of the apparent integrated absorption intensities, the molecular weight determination is the largest source of possible error in the analysis of compound types by infrared spectroscopy.

The results of the infrared analysis are presented in Table VI. These results show that carboxylic acids and phenols are found only in the acid concentrates. Carboxylic acids are concentrated in the polar acid subfractions III and IV while phenols are concentrated in subfraction II. Carbazoles, ketones, and amides are found in all three major nonhydrocarbon fractions. The appearance of the same compound type in several fractions may arise from differences in acidity or basicity that are caused by the hydrocarbon portion of the molecule. Multifunctionality could also be a factor in the distribution of compound types among the fractions. The 1695 cm$^{-1}$ band was assigned to ketones on the basis of work

**of P. R. Spring Heteroatomic Concentrates**

| | | Compound Type (wt % of fraction) | | | | |
|---|---|---|---|---|---|---|
| Phenol | Carbazole | Carbox-ylic Acid | Ketone | Amide | Sulfoxide | Total |
| 18 | 68 | 8 | 100 | 12 | | 206 |
| 40 | 36 | 56 | | 66 | | 198 |
| 12 | 14 | 120 | | 20 | | 166 |
| 8 | 4 | 78 | | 18 | | 108 |
| | | | 96 | 8 | 25 | 129 |
| | | | 82 | 36 | 21 | 139 |
| | | | 43 | 9 | 8 | 60 |
| | 13 | | 63 | 21 | 20 | 117 |
| | 23 | | 131 | 11 | 25 | 190 |
| 2.6 | 12.3 | 9.8 | 30.0 | 8.2 | 5.6 | 68.5 |

done by Dorrence *et al.* (20) where the 1695 cm$^{-1}$ band present in asphalts was reducible by sodium borohydride. Ketones are reduced by this mild reducing agent while amides are not.

Sulfoxides were found in the weak bases and the neutral Lewis bases. The presence of sulfoxides in the bases is explained by their weakly basic character (21). The presence of sulfoxides in the neutral Lewis bases probably results because the sulfur–oxygen bond is extremely polar and may complex with ferric chloride. Elemental analysis of the sub-fractions (discussed later) suggests that the sulfoxides are concentrated by the separation scheme and are not oxidation products that occur after the separation. The totals which are in excess of 100% reflect the degree of multifunctionality as well as possible errors in VPO molecular weight.

The three base fractions were titrated for basic nitrogen to determine the average strength of the bases and the amount of titratable nitrogen present. All titratable material was assumed to be nitrogen bases, although sulfoxides are present and may titrate as weak bases (18, 21). The results are given in Table VII. The first two base fractions contain

### Table VII. Potentiometric Titration of Bases

| Base Fraction | Strong Bases | | Very Weak Bases | | Total N (Analysis) |
|---|---|---|---|---|---|
| | HNP (mv) | % N (tit.) | HNP (mv) | % N (tit.) | |
| I | | | 483 | 1.12 | 1.6 |
| II | | | 429 | 1.42 | 2.1 |
| III | 231 | 1.65 | 466 | 0.16 | 2.7 |

only very weak bases. These bases titrate with a half-neutralization potential (HNP) roughly equivalent to model amides and sulfoxides. As expected from the separation procedure, bases II are generally stronger than bases I. Bases III, which comprise over 70% of the total bases, contains high concentrations of strong bases; essentially no weak bases are present in this fraction. An HNP of 231 could result from alkyl pyridines or quinolines. A tentative identification of condensed benzologs of pyridine has been made in this P. R. Spring bitumen by separation by thin layer chromatography and fluorescence spectroscopy.

Analysis of the hydrocarbons was attempted by mass spectroscopy. Analysis of the saturated hydrocarbons was obtained using a modification of the mass spectral method (19) originally proposed by Hood and O'Neal. This method allows the determination of saturated hydrocarbons according to number of rings. Table VIII lists the results of this analysis. The data show that over 60% of the saturates are two- and three-ring compounds. The low percentage of no-ring compounds suggests either

**Table VIII.   Group-type Analysis of P.R. Spring Saturated Hydrocarbons**

| Number of Rings | Wt % of Saturates |
|---|---|
| 0 | 7.1 |
| 1 | 12.3 |
| 2 | 29.4 |
| 3 | 31.5 |
| 4 | 14.1 |
| 5 | 4.4 |
| 6 | 1.3 |
| Monoaromatics | 0 |

an immature oil or one which has been biodegraded. Analysis of aromatic hydrocarbons was not obtainable because of the wide molecular weight range represented.

**The Nitrogen and Sulfur Analysis of Defined Fractions.** Nitrogen and sulfur analysis for the subfractions are given in Table IX. For the acid fractions, the nitrogen is concentrated in the weak acids. This is consistent with the infrared analysis that showed these fractions to contain predominantly amides and carbazoles. The sulfur in the acid concentrates is randomly distributed and is probably of a thiophene or sulfide type. There is no evidence for the presence of appreciable quantities of sulfur–oxygen compounds such as sulfoxides, sulfones, or sulfonic acids in the acid concentrates.

The nitrogen content in the base fractions increases with increasing polarity. If monofunctionality is assumed, this would indicate that the molecular weight decreases as the polarity increases. This is contrary to the VPO molecular weight data, which again indicates substantial

**Table IX.   Nitrogen and Sulfur Analysis of Concentrates**

| Concentrate | Wt % | | |
|---|---|---|---|
| | Bitumen | Nitrogen | Sulfur |
| Acids I | 4.8 | 2.0 | 0.8 |
| Acids II | 2.2 | 2.7 | 0.8 |
| Acids III | 4.0 | 1.3 | 0.9 |
| Acids IV | 4.4 | 1.8 | 0.7 |
| Bases I | 2.0 | 1.6 | 2.2 |
| Bases II | 1.5 | 2.1 | 2.2 |
| Bases III | 8.8 | 2.7 | 0.8 |
| Neutral Lewis bases I | 10.3 | 1.0 | 1.6 |
| Neutral Lewis bases II | 8.2 | 1.4 | 1.3 |
| Hydrocarbons | 50.6 | 0.1 | 0.2 |
| Totals (wt % of bitumen) | 96.8 | 0.86 | 0.64 |

error in the molecular weight determination of polar compounds. Sulfur is concentrated in the weak base fractions. These fractions have been shown by IR to contain substantial quantities of sulfoxides.

The neutral Lewis bases are surprisingly low in nitrogen content. Because of the neutral character of these molecules the VPO molecular weight of the neutral Lewis bases I of 1000 may, in this case, accurately reflect the average molecular weight. The neutral Lewis bases contain sulfoxides that the acids do not, which could account for the higher percentage of sulfur in the neutral Lewis bases.

Of the total nitrogen present in the bitumen, only 5.9% of it is found in the hydrocarbon concentrates, which is essentially complete removal of nitrogen from the hydrocarbons. The comparable percentage for sulfur is 15.9%; a separation of sulfur compounds analogous to that for the nitrogen compounds has not been developed. Observation of the distribution of the sulfur among the concentrates, in conjunction with the IR analysis for sulfoxides, leads to the speculation that sulfide- and thiophene-type sulfur is randomly distributed among molecules containing other heteroatoms. For the heteroatomic concentrates, the percentage of sulfur represented by these types is about 0.8%. The sulfur content in the concentrates that is in excess of 0.8% correlates with the infrared analysis for sulfoxides, indicating that the separation procedure is sensitive to sulfoxides but not to sulfide- or thiophene-type sulfur.

## Conclusions

The results of this work show that a separation according to functional groups has been made. The separation serves two important purposes. First, it allows a more precise and accurate analysis by conventional spectral and analytical techniques. This is accomplished primarily because resolution and intensity of spectral bands are greatly enhanced and a more accurate determination of the molecular weight of a specific type can be made. It is also important that definition by the separation scheme allows the accurate assignment of bands, particularly in the carbonyl region. Second, the separation provides simplified subfractions that are amenable to the determination of structure, properties, and reactivity of the compound types present. McKay et al. (12) have shown that detailed structural information can be obtained on petroleum acids derived in this fashion. Structural analysis of petroleum saturated and aromatic hydrocarbons derived by this separation procedure has been accomplished by Dooley et al. (22). The detailed analysis of petroleum samples has been limited to high boiling crude oil distillates, and the application of these techniques to petroleum residues and tar sand bitumens must recognize additional problems resulting from decreased solubility and volatility. However, the results of this analysis demonstrate that these procedures can be generally applied to analyze heavy bitumens and residues.

## Acknowledgments

Many stimulating discussions and helpful suggestions from J. F. McKay are gratefully acknowledged. Appreciation is expressed to Texaco, Inc., for supplying the P. R. Spring core sample.

## Literature Cited

1. Wood, R. E., Ritzma, H. R., "Analysis of Oil Extracted from Oil-Impregnated Sandstone Deposits in Utah," *Utah Geol. Mineral. Surv., Spec. Stud.* (1972) **39**.
2. Gwynn, J. W., "Instrumental Analysis of Tars and Their Correlations in Oil-Impregnated Sandstone Beds, Uintah and Grand Counties, Utah," *Utah Geol. Mineral. Surv., Spec. Stud.* (1971) **37**.
3. Camp, F. W., "The Tar Sands of Alberta, Canada," 2nd ed., 77 pp., Cameron Engineers, Denver, Colo., 1974.
4. Speight, J. G., *Fuel* (1970) **49**, 76.
5. Jones, J. H., Moote, T. P., *Am. Chem. Soc., Div. Petrol. Chem., Preprint* (1963) **8** (2), A-29.
6. Speight, J. G., *Fuel* (1971) **50**, 102.
7. Augston, D. M., George, A. E., Montgomery, D. S., Smiley, G. T., Sawatzky, H., *Am. Chem. Soc., Div. Fuel Chem., Preprints* (1974) **19** (2), 202.
8. Goodspeed, F. E., Montgomery, D. S., *Mines Branch Research Report, R98*, 25 pp. Dept. of Mines and Technical Survey, Ottawa, Canada, Apr., 1962.
9. Boyd, M. L., Montgomery, D. S., *Mines Branch Research Report*, **R78**, 88 pp., Dept. of Mines and Technical Surveys, Ottawa, Canada, Dec., 1961.
10. Jewell, D. M., Weber, J. H., Bunger, J. W., Plancher, H., Latham, D. R., *Anal. Chem.* (1972) **44**, 1391.
11. McKay, J. F., Jewell, D. M., Latham, D. R., *Sep. Sci.* (1972) **7**, 361.
12. McKay, J. F., Cogswell, T. E., Latham, D. R., *Fuel* (1975) **54**, 50.
13. Latham, D. R., Haines, W. E., *Fuel* (1973) **18** (3), 567.
14. Haines, W. E., Ward, C. C., Sugihara, J. M., *Proc. Am. Pet. Inst., Div. Refin.* (May 1971) 375.
15. Poulson, R. E., Jensen, H. B., Duvall, J. J., Harris, F. L., Morandi, J. R., *Anal. Instrum.* (1972) **10**, 193.
16. Dorrence, S. M., personal communication.
17. Petersen, J. C., personal communication.
18. Buell, B. E., *Anal. Chem.* (1967) **39**, 756.
19. Standard Method for High Ionizing Voltage Mass Spectrometric Analysis of Gas-Oil Saturate Fractions, *Book ASTM Stand.* **D2786-71** (1971), 1019.
20. Dorrence, S. M., Barbour, F. A., Petersen, J. C., *Anal. Chem.* (1974) **46**, 2242.

21. Okuno, I., Latham, D. R., Haines, W. E., *Anal. Chem.* (1967) **39** (14), 1830.
22. Dooley, J. E., Thompson, C. J., Hirsch, D. E., Ward, C. C., *Hydrocarbon Process.* (1974) **53** (4), 93.

RECEIVED December 16, 1974. The work reported in this paper was performed under a cooperative agreement between the Bureau of Mines, U.S. Department of the Interior, and the University of Wyoming. Mention of specific brand names or models of equipment is made for information only and does not imply endorsement by the Bureau of Mines.

# 11

# Feasibility Studies of a Biochemical Desulfurization Method

A. J. DAVIS, III and T. F. YEN

Departments of Biological Sciences and Chemical Engineering, University of Southern California, University Park, Los Angeles, Calif. 90007

*The microorganism,* Thiobacillus thiooxidans, *is a possible desulfurizing agent of organically bound sulfur with an efficiency greater than 20%. It oxidizes inorganic sulfur to sulfate, producing sulfuric acid as a by-product. In this study,* T. thiooxidans *was grown on tertiary butylsulfides (mono-, di-, and poly-) for 24 days. During this time, growth of the organism was followed by a decrease in pH and an increase in sulfate production, gravimetrically assayed for in the barium salt form. There was significant growth on the tertiary butyldisulfide and the tertiary butylpolysulfide. Significant growth was not reported on the tertiary butylmonosulfide.*

The removal of sulfur from high sulfur containing petroleum has long presented a problem. With the depletion of so-called "sweet" oil (*i.e.*, oil low in sulfur content), the problem has escalated into major proportions to contend with a clean and livable environment. The development of practical methods of decreasing or eliminating the sulfur content of petroleum has proved repeatedly to be an economical obstacle to industry. The amount of dollars spent on corrosion technology alone is phenomenal (*1*).

Rudimentary investigations of microbial desulfurization have received little attention in the literature at this time (*2*). One successful example of desulfurization is the removal of pyrite from coal by *Thiobacillus sp.* and *Ferrobaccus sp.* (*3*). While studies of the complex hydrocarbon–sulfur systems are of great value, being closer to *in situ* reality, investigation of a defined system should form the foundation of these more detailed studies.

Our present study was intended to explore the ability of a micro-organism as a possible desulfurizing agent. To this end, a pure, defined system was used that would give some idea as to the ease with which such an agent could abstract organically bound sulfur. While this program is preliminary, future investigations may lead to a feasible industrial application.

## Materials and Methods

**Media and Culture.** The strain of *Thiobacillus thiooxidans* used in these experiments was originally obtained from the National Types Culture Collection. Inoculate substrain was obtained by further culture in this laboratory (4). Fernbach culture flasks (2.8-l. borosilicate glass) containing 1.0 l. of Waksman's medium were prepared in duplicate. The medium contained the following concentrations of salt in g/l. (all Mallinckrodt "AR" grade): $(NH_4)_2SO_4$, 0.20; $KH_2PO_4$, 3.00; $MgSO_4 \cdot 7H_2O$, 0.50; $CaCl_2 \cdot 2H_2O$, 0.25. The medium was prepared with distilled water. In place of the elemental sulfur substrate of Waksman's medium, different symmetric organic sulfides of the tertiary butyl class with a normalized sulfur equivalent of 10.0 g/l. were used. To provide a greater surface area, the normally viscous sulfides were emulsified with Triton X-100 surfactant (J. T. Baker Chemical Co.). The first series of flasks received 49.0 ml *tert*-butyl sulfide and one drop Triton X-100 each. The second series of flasks received 28.0 ml *tert*-butyl disulfide and one drop Triton X-100 each. The third series of flasks received 17.0 ml *tert*-butyl polysulfide and two drops Triton X-100 each. In addition, flasks of elemental sulfur and flasks of organic sulfur substrates were prepared to act as controls. The pH of the media was adjusted to 3.5 with 1.0N $H_3PO_4$ (phosphoric acid) (Mallinckrodt "AR" grade), and sterilization was accomplished by autoclaving or by membrane filtration. After sterilization all flasks containing sulfur or organic sulfur compounds, except the organic sulfur control flasks, were inoculated with 10.0 cc *T. thioxidans* suspensions. All flasks were incubated at room temperature.

**Sulfide Samples.** The *tert*-butyl sulfide and the *tert*-butyl disulfide used in these experiments were obtained from the Aldrich Chemical Co. The purity of the monosulfide (MW 146.30) was in excess of 97%. The technical grade disulfide (MW 178.36) was 38% pure and contained 62% tertiary butyl trisulfide.

The *tert*-butyl polysulfide used was kindly supplied by Phillips Petroleum Co. The polysulfide (MW 190.00) was greater than 90% pure. The polysulfide contained an average of four and five sulfur atoms per molecule. No pretreatment of the sulfides was performed.

**Analytical Methods.** Insoluble barium sulfate precipitate, one of the two criteria by which growth of the bacteria was established was determined by standard gravimetric methods. The amount of barium sulfate is directly proportional to the amount of sulfide that has been oxidized by the sulfur bacteria.

Hydrogen ion concentration (pH) was the second criterion by which growth of the bacilli was established. Sulfuric acid is a natural metabolic by-product of sulfur oxidation by the acidophillic *Thiobacillus thiooxidans* (5). As sulfur is used, acid is built-up in the medium thus lowering the pH. Studies in this laboratory have shown that the bacteria grow well in a pH as low as 0.5.

Samples for the barium sulfate determinations were prepared by removing a 5.0 ml aliquot portion from each flask and then centrifuging them on an Adams centrifuge for 15 min at 3000 rpm. To the supernatant of each was added 2 drops concentrated hydrochloric acid, an acid buffer, and 11 drops of a saturated solution of barium chloride (Mallinckrodt "AR" grade). The precipitate was filtered on a Millipore apparatus (borosilicate glass), washed with distilled water, and then allowed to dry three days in an evaculated dessicator. The samples were then ignited, and the weight of the barium sulfate established. The controls were handled in the same manner. The amount of barium sulfate produced from the organic sulfides less the controls was averaged and then plotted. The pH was read directly with a Beckman Zeromatic, SS-3 pH meter.

*Procedure*

Modified Waksman's medium was prepared in a Fernbach culture flask. An amount of organic sulfide normalized to an equivalent sulfur content of the standard medium (10 g/l.) was added followed by an emulsifier. The medium was then autoclaved for 30 min at 15 psi (121°C), or, as with the polysulfide and elemental sulfur, sterilization was achieved by membrane filtration. Upon cooling, the medium was inoculated with 10 cc of the pure strain of *Thiobacillus thiooxidans*. The culture's initial pH value was read, and an initial gravimetric sulfate assay was performed. Thereafter, pH and sulfate values were determined at two-day intervals for 25 days.

*Results*

The model was designed to study the ability of sulfur-oxidizing bacteria to use the organically bound sulfur as substrate. The symmetry of the sulfides provided an insight into the sulfur abstracting prowess of *T. thiooxidans*.

Figure 1 shows that the oxidizing potential of the organism is enhanced by the availability of unshielded sulfur in the molecular structure. The sulfur–sulfur bonds of the di- and polysulfides are easily disrupted; but access to the carbon–sulfur–carbon bond of the monosulfide seems

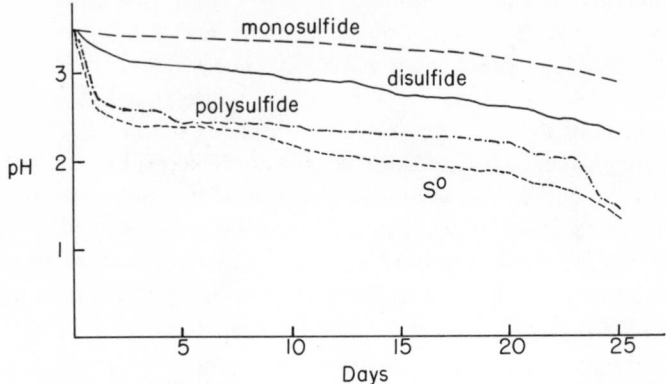

*Figure 1.   Conversion of organic sulfide into sulfate in 25-day period*

*The amount of sulfate ion represents the difference of that produced in the inoculated flasks containing different organic sulfide and that in the controls containing the same organic sulfide without inoculation.   —·—, di-tert-butyl polysulfide;  — — —, di-tert-butyl disulfide;  ——, di-tert-butyl monosulfide;  — – —, control on elemental sulfur.*

questionable. While some sulfate ion is produced by the organism on the monosulfide, it is not known how much of the sulfate arises from residual sulfur previously incorporated by the bacteria. The values of pH seem to bear this out (Figure 2). Based on sulfate ion and pH values, however, it appears that the organism is growing on the di- and polysulfides.

### Discussion

The polysulfide and elemental sulfur were unstable to autoclaving. This fact necessitated the use of a membrane to achieve sterility in those cases. The surfactant not only provided greater surface area but also minimized the amount of sulfide in contact with atmospheric oxygen since the sulfide was generally dispersed throughout the medium. The fate of the non-sulfur-containing hydrocarbons, while of real interest, is beyond the scope of this present work.

This study indicates that the sulfur of our sulfide samples was susceptible to bacterial attack in two of the three cases. The aliphatic sulfides could be ranked in the order of their ease of oxidation as: di-*tert*-butyl polysulfide > di-*tert*-butyl disulfide > di-*tert*-butyl sulfide.

Seemingly, *T. thiooxidans* is able to attack the sulfur–sulfur bond quite readily and the sulfur–carbon bond with some difficulty, if at all. Since a large portion of the sulfur compounds in petroleum are of the monosulfide-aromatic type, more attention is paid to this facet.

The simplest scheme of the bacterial oxidation would be:

$$S^{-2}\ T.\ \xrightarrow[\ \ \ \ \ \ \ \ \ \ ]{T.\ thiooxidans}\ SO_4^{-2}$$

A few mechanisms were postulated for this reaction (*6*). No matter what mechanism is considered, the oxidation of elemental sulfur or thiosulfate is accompanied by reductive cleavage of the sulfur–sulfur bridges. In the case of sulfur, the intermediate involved is a cyclic form of sulfur, probably $S_8$, although there is little difference observed for different allotrophic forms of sulfur such as rhombic, precipitated, and amphorous. These cyclic sulfides form the basis of polysulfanes and polythionates which could be metabolized readily by *Thiobacilli* (as shown in Figure 3). Actually, sulfur oxidation begins with its reduction, in which the gluathione–sulfhydryl groups located near the cell surface take part:

$$\text{NAD-H} + \text{G-S-S-G} \rightarrow \text{NAD} + 2\text{G-SH}$$

$$\text{S}° + 2\text{G-SH} \rightarrow \text{H}_2\text{S} + \text{G-S-S-G}$$

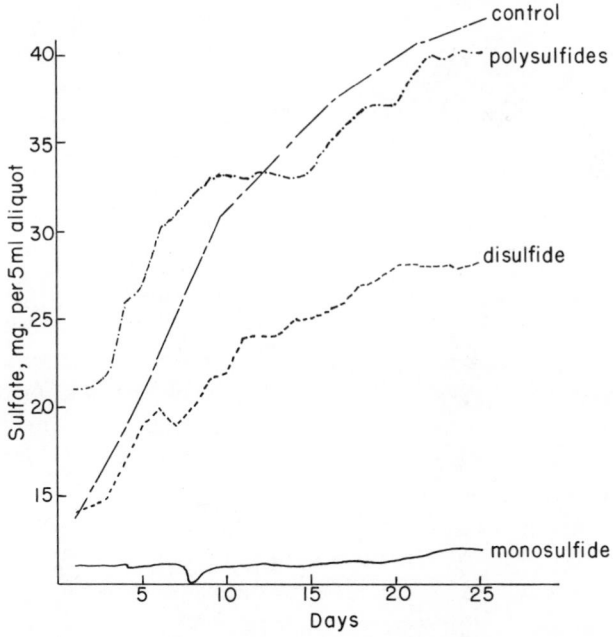

*Figure 2. Variation of pH after inoculation of sulfur bacteria.* — — —, *di-tert-butyl monosulfide;* ———, *di-tert-butyl disulfide;* — · —, *di-tert-butyl polysulfide;* – – –, *control of elemental sulfur.*

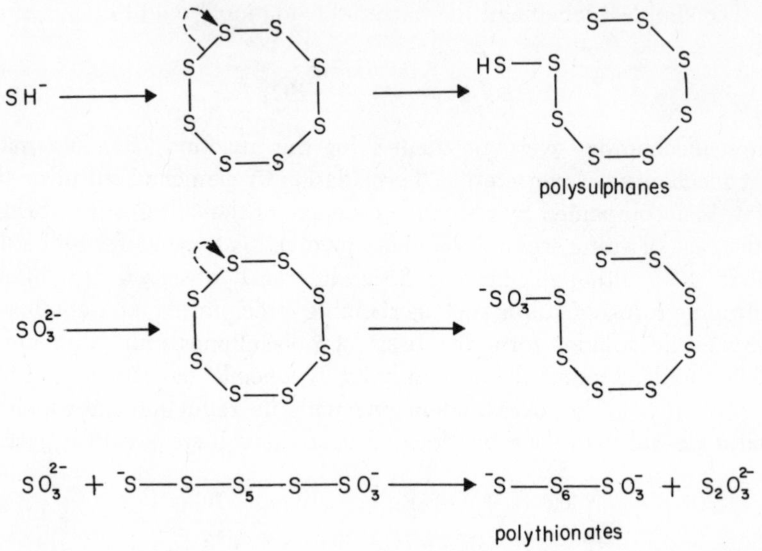

$$SO_3^{2-} + {}^-S\!-\!S\!-\!S_5\!-\!S\!-\!SO_3^- \longrightarrow {}^-S\!-\!S_6\!-\!SO_3^- + S_2O_3^{2-}$$

polythionates

*Figure 3. Reductive cleavage of S–S bridges of polysulfanes and polythionates*

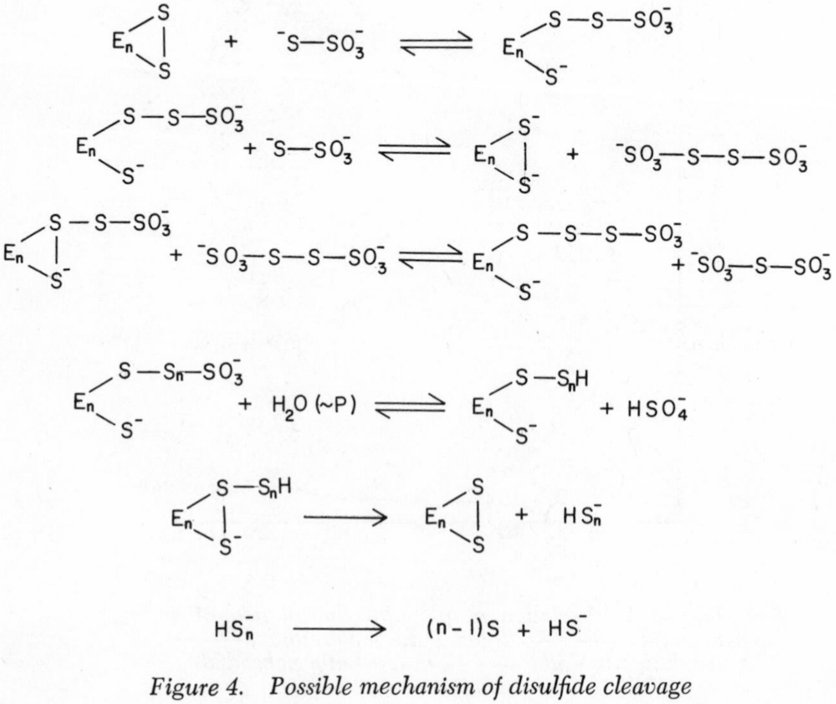

$$HS_n^- \longrightarrow (n-1)S + HS^-$$

*Figure 4. Possible mechanism of disulfide cleavage*

It is thought that the SH⁻ of the glutathione–sulfhydryl group acts as the functional agent in the reductive cleavage of the sulfur.

For thiosulfate oxidation it is hypothesized that an enzyme system on the cell surface initiates the formation of polythiosulfonic acid and consequently splits the terminal $SO_3^-$ as sulfate. This process may evolve the intermediate of mixed anhydride -S-O-$PO_4^-$ from phosphorylation (Figure 4).

The overall reaction is one of electron transfer in the enzyme system:

$$E_n \overset{S}{\underset{S}{\bigg\langle}} \quad \overset{+2e}{\underset{-2e}{\longrightarrow}} \quad E_n \overset{S^-}{\underset{S^-}{\bigg\langle}}$$

The critical step in the use of multiple sulfur linkages is the availability of HS⁻ from the reduced enzyme system. The HS⁻ could possibly attack the cyclic or the acyclic multiple sulfur linkages through a nucleophilic mechanism. The cleaved sulfide linkages undergo oxidation to sulfite or thiosulfate. This seems a plausible pathway for the oxidation of the sulfur compounds in the present study.

Finally, removal of the sulfate ion creates a new contamination problem. In connection with this, sulfate reduction bacteria for the complete removal of sulfur are being examined. The conversion of the organic sulfide into inorganic sulfide will be one future objective of this study.

*Acknowledgment*

The authors thank Judith Higa for technical help.

*Literature Cited*

1. Bertness, T. A., Mobil Oil Corp., personal communication, 1973.
2. Isenberg, D. L., "Microbial Desulfurization of Petroleum," Ph.D. Dissertation, Louisiana State University, 1961.
3. Silverman, M. P., Rogoff, M. H., Wender, I., "Removal of Pyritic Sulfur from Coal by Bacterial Action," *Fuel* (1963) **42**, 113.
4. Findley, J. E., Appleman, M. D., Yen, T. F., "Abstracts of Papers," 166th National Meeting, ACS, Div. of Microbial Chemistry and Technology, Chicago, 1973.
5. Roy, A. S., Trudinger, P. A., "The Biochemistry of Inorganic Compounds of Sulfur," p. 207, Cambridge University, 1970.
6. Sokolova, G. A., Karavaiko, G. I., "Physiology and Geochemical Activity of *Thiobacilli*," Israel Program for Scientific Translations, p. 48, 1968.

RECEIVED December 26, 1974. This work is supported by NSF Grant No. GI-35683, AER-74-23797, and ACS-PRF Grant No. 6272-AC2.

# 12

# Chemical Modification of Bitumen Heavy Ends and Their Non-Fuel Uses

SPEROS E. MOSCHOPEDIS and JAMES G. SPEIGHT

Fuel Sciences Division, Alberta Research Council, 11315—87th Avenue, Edmonton, Alberta, Canada T6G 2C2

*Bitumen asphaltenes undergo a variety of simple chemical conversions. For example, asphaltenes can be oxidized, sulfonated, sulfomethylated, halogenated, and phosphorylated. The net result is the introduction of functional entities into the asphaltene structure which confer interesting properties on the products for which a variety of uses are proposed.*

The development of the Athabasca tar sands has become one of the major advances of the petroleum industry. However, the production of materials from the bitumen is only just beginning, and with the introduction of gas turbine, electric, propane autos, and the like, the future of the tar sands may not lie in the production of gasoline but in the use of bitumen as a chemical raw material. In this paper, we report some simple chemical conversions that may be used to introduce functional groups into the bitumen heavy ends and some preliminary investigations on the uses of the products. For convenience, we describe the reactions of the asphaltenes—a fraction considered somewhat useless (except for fuel purposes) by many workers.

## Chemical Reactions

As a result of structural studies, it is evident that petroleum asphaltenes are agglomerations of compounds of a condensed aromatic type (1, 2). Thus, it is not surprising that asphaltenes undergo a wide range of chemical and physical interactions based not only on their condensed aromatic structure but also on the attending alkyl and naphthenic moieties. In the following discussion, we again rely heavily on the evidence accumulated, for the major part, in our own laboratries, which has either been published elsewhere or is in preparation.

**Oxidation.** Oxidation of Athabasca asphaltenes with a series of common oxidizing agents, such as acidic and alkaline peroxide, acidic dichromate, and alkaline permanganate (20 hr/20°C), is a slow process. Less than 10% of the product is soluble in alkali after treatment with the oxidant for *ca.* 30 hr (*4*). It is evident, however, from the elemental ratios in the products that some oxidation has occurred. Moreover, the occurrence of a broad band centered at 3420 cm$^{-1}$ and a weaker band at 1710 cm$^{-1}$ indicates the formation of phenolic and carboxyl groups on the asphaltene molecules. The H/C ratios of the partially oxidized products indicate that there are three predominant oxidation routes, notably the oxidation of naphthene moieties to aromatics and active methylene groups to ketones which would reduce the H/C ratios and more severe oxidation of naphthene and aromatic functions resulting in partial degradation of these systems to carboxylic acid functions. Investigation of the partially oxidized materials by proton magnetic resonance spectroscopy (*4*) provides information about the structural changes which occur during the oxidation processes. In this case there was a decrease in size of the condensed aromatic sheet, a slight decrease in the degree of substitution of the sheet, and a decrease in average chain lengths of the alkyl substituents. Use of a more severe oxidizing agent, such as concentrated nitric acid, caused good conversion of the asphaltenes to water- and alkali-soluble materials (*5*).

**Sulfonation and Sulfomethylation.** Sulfomethylation and/or sulfonation of the asphaltenes are not feasible, but oxidation of the asphaltenes produces the necessary functional groups which allow sulfomethylation and sulfonation (*5*). Confirmation of this can be obtained from three sources:

1. Overall increases (up to 40%) in the sulfur contents of the products;

2. The appearance of a new infrared absorption band at 1030 cm$^{-1}$ attributable to the presence of sulfonic acid group(s) in the molecule(s) (*6*)

3. The water solubility of the products—a characteristic of this type of material (*7, 8*).

These sulfomethylated and sulfonated oxidized asphaltenes even remain in solution after acidification with 5% aqueous hydrochloric acid to a pH of 2.5–3.0 while the parent-oxidized asphaltenes can be precipitated from alkaline solution by acidification to a pH of 6.5.

The facile sulfomethylation reaction indicates the presence in the starting materials of reactive sites ortho or para to a phenolic hydroxyl group, *e.g.;*

while the comparative ease of sulfonation suggests the presence of quinoid structures in the oxidized materials, *e.g.;*

Alternatively, active methylene groups in the starting materials would facilitate sulfonation by:

since such groups have been known to remain intact after prolonged oxidation (*9*).

**Halogenation.** Halogenation of the asphaltenes—by addition of the halogen to a solution of the asphaltenes in refluxing carbon tetrachloride —occurs readily to afford the corresponding halo derivatives (*10*). The physical properties of the halogenated materials are markedly different from those of the parent asphaltenes. For example, the unreacted asphaltenes are dark brown, amorphous, and readily soluble in benzene, nitrobenzene, and carbon tetrachloride, but the products are black, shiny, and only sparingly soluble, if at all, in these solvents.

The individual halogen reactions differ from one another in several other features. During chlorination the asphaltenes stop taking up chlorine after 4 hr. Analytical data indicate that more than 37% of the total chlorine in the final product is introduced during the first 0.5 hr, reaching the maximum after 4 hr. Furthermore, the $(H + Cl)/C$ ratio in the chlorinated materials remains at 1.22 during the first 2 hr of chlorination. H/C ratio in the parent asphaltenes is 1.22. This is interpreted as substitution of hydrogen atoms by chlorine in the alkyl moieties of the asphaltenes while the condensed aromatic sheets remain unaltered since substitution of aryl hydrogens only appears to occur readily in the presence of a suitable catalyst, *e.g.*, ferric chloride, or at elevated temperatures. It is only after more or less complete reaction of the alkyl chains that addition to the aromatic rings occurs as evidenced by the increased $(H + Cl)/C$ ratios in the final stages of chlorination.

Bromine uptake by the asphaltenes is also complete in a comparatively short time ( < 8 hr). However, in contrast to the chlorinated products, the $(H + halogen)/C$ ratio remains fairly constant (1.23 and 1.21 in the bromo–asphaltenes compared with 1.22 in the unreacted asphaltenes) over the prolonged periods (up to 24 hr) of the bromination.

Iodination of the asphaltenes is different insofar as a considerable portion of the iodine, recorded initially as iodine uptake, can be removed by extraction with ether or with ethanol while there is very little weight loss after prolonged maintenance of the material in a high vacuum. The net result is the formation of a product with a $(H + I)/C$ ratio of 1.24 after 8 hr reaction, but a longer reaction period affords a product having a $(H + I)/C$ ratio of 1.17. This latter result may be interpreted as iodination of the alkyl or naphthenic moieties of the asphaltenes with subsequent elimination of hydrogen iodide. Alternatively, oxidation of naphthenic moieties to aromatics or oxidative coupling of asphaltene nuclei would also account for lower $(H + I)/C$ ratios. In fact, this latter phenomenon could account, in part, for the insolubility of the products in solvents which are normally excellent for dissolving the unchanged asphaltenes. Halogenation of the asphaltenes can also be achieved by use of sulphonyl chloride, iodine monochloride, and *N*-bromosuccinimide, or *via* the Gomberg reaction (*11*) whereby products similar to those described above are produced.

It is not possible to dissolve the asphaltenes in water by treatment of the halo derivatives with aqueous sodium hydroxide or with aqueous sodium sulfite (*11*). The hydrolyzed products remained insoluble even in a strongly alkaline solution. The decreased $(H + Cl)/C$ ratios and the increased $O/C$ ratios of the products relative to those of the parent haloasphaltenes indicate that partial reaction occurs. The infrared spectra of the products showed a broad band centered at 3450 cm$^{-1}$, a region assigned to the presence of hydroxyl groups in the products, but it was not possible to establish conclusively the presence of sulfonic acid group(s) in the product from the sodium sulfite reaction by assignment of infrared absorption bands to this particular group.

**Reactions with Metal Salts.** Interactions of asphaltenes with the metal chlorides yield products containing organically bound chlorine, but the analytical data indicate dehydrogenation processes occurring simultaneously (*12*). There is, of course, no clear way by which the extent of the dehydrogenation can be estimated, but it may be a dehydrogenation condensation rather than elimination with olefin formation. Infrared spectroscopy did not show any bands that could be unequivocally assigned to C=C bond vibrations, nor did proton magnetic resonance (pmr) spectroscopy show any olefinic protons. The mode of dehydrogenation is assumed to involve predominantly inter- or intramolecular condensation reactions because the solubilities and apparent complexities of the products varied markedly from those of the starting materials. These differences could not be attributed wholly to the incorporation of chlorine atoms into the constituents of the asphaltenes or heavy oil. Indeed, the data accumulated indicate a condensation dehydrogenation or, in part, loss of alkyl substituents as, for example, lower molecular weight hydrocarbons during the reactions. As an illustration of the former, the cokes produced during the thermal cracking (450°C) of the asphaltenes or heavy oil have H/C ratios in the range 0.59–0.77 (16) while the majority of the insoluble materials produced in the asphaltene/metal chloride reactions have only slightly higher $(H + Cl)/C$ ratios (0.88–1.10).

**Reactions with Sulfur and Oxygen.** Reaction of asphaltenes with sulfur and oxygen have also received some attention and have yielded interesting results (13). For example, treatment of the asphaltenes with oxygen or with sulfur at 150°–250°C yields a condensed aromatic product ($H/C = 0.97$, cf. $H/C_{asphaltenes} = 1.20$) containing very little additional oxygen or sulfur. The predominant reaction appears to be condensation between the aromatic and aliphatic moieties of the asphaltenes effected by the elemental oxygen and sulfur, which are in turn converted to hydrogen sulfide and water. With oxygen, condensation appears to precede in preference to molecular degradation, but prolonged reaction times yield lower molecular weight products. Treatment of the condensed products at 200°–300°C for 1–5 hr again yields good grade cokes ($H/C = 0.54$–0.56). In all instances the final products contain only very low amounts of elements other than carbon and hydrogen (i.e., $\Sigma NOS < 5\%$ w/w)—a desirable property of good grade coke.

**Phosphorylation.** Attempts to phosphorylate asphaltenes with phosphoric acid, phosphorous trichloride, or phosphorous oxychloride were partially successful insofar as it is possible to introduce up to 3% w/w phosphorous into the asphaltenes (14). However, application of these same reagents to oxidized asphaltenes increases the phosporous quite markedly to 10% w/w in the products (14). Subsequent reaction of the phosphorous-containing products is necessary to counteract the acidity of the phosphorous moieties.

**Hydrogenation.** Montgomery and his co-workers (15, 16) studied the reduction products of tar sand asphaltenes and noted that considerable amounts of hydrogen could be added to the asphaltenes while some sulfur was removed. Other studies of the hydrogen processes have investigated mainly the effects of additional hydrogen on the cracking. Increased yields of paraffins were observed (3).

**Miscellaneous Chemical Conversions.** Other reactions of the asphaltenes have also been performed, but the emphasis has been mainly on the formation of more condensed materials to produce good grade cokes. For example, thermal treatment of the halogenated derivatives affords aromatic cokes containing less than 1% w/w halogen and $H/C = 0.58$. Coke from the thermal conversion of asphaltenes at $\sim 460°C$ has $H/C = 0.77$ (13, 17). Other investigations (13) also show that treatment of the halo-asphaltenes with suitable metal catalysts, *e.g.*, copper at 200°–300°C /1–5 hr or sodium at 80°–110°C/1–5 hr, yield aromatic ($H/C = 0.55$–0.86, respectively) coke–like materials having 0.5–3% w/w halogen. Residual halogen may finally be removed by treatment at 300°C for 5 hr. These investigations have many implications to the petroleum industry. Perhaps the most significant two are the comparatively lower temperatures required for coking caused by the presence of bonds in the asphaltenes which are relatively labile and the use of oxygen or sulfur as condensing and aromatizing agents without being significantly incorporated into the ensuing coke. Further work on the catalytic effects of other readily available inorganic materials on the coking process are still under investigation, and all of the data will be reported in detail.

Other chemical modifications pursued in our laboratories include metallation of the asphaltenes or halo-asphaltenes using metal or metalloorganics followed by, for example, carboxylation to the end product. Interaction of the asphaltenes with *m*-dinitrobenzene affords an oxygen-enriched material which, when treated with hydroxylamine or another amine yields materials containing extra nitrogen. Similarly, reaction of the asphaltenes with maleic anhydride and subsequent hydrolysis yields product bearing carboxylic acid functions.

### Applications

The traditional uses of petroleum involve the derivation of chemicals from the oil during a refinery operation. It is usual that at some stage during the operation asphaltenes are removed as a sludge either to be discarded later or to be used as a fuel source. We have attempted to show how the asphaltenes may be regarded as chemical entities which are able to undergo a variety of chemical or physical conversions to more useful materials. The overall effect of these modifications is the production of materials which afford either good grade aromatic cokes or products bearing functional groups which may be used as a non-fuel material. To date, our main tests have centered around the sulfonated and sulfo-

Table I.  Thinning Properties and Comparison of

| Designation | lb/bbl | $NaOH$ lb/bbl | $CaO$ lb/bbl | pH |
|---|---|---|---|---|
| Base mud[b] | | 2 | 5 | 12.4 |
| Sulfomethylated | 2 | — | — | 11.6 |
| asphaltene | 4 | — | — | 11.5 |
| | 6 | — | — | 11.5 |
| Sulfonated asphaltenes | 2 | — | — | 11.5 |
| | 4 | — | — | 11.5 |
| | 6 | — | — | 11.5 |
| UNI–CAL | 2 | — | — | 12.6 |
| | 4 | — | — | 12.6 |
| | 6 | — | — | 12.6 |
| Spercene | 2 | — | — | 12.6 |
| | 4 | — | — | 12.6 |
| | 6 | — | — | 12.6 |
| Peltex | 2 | — | — | 12.5 |
| | 4 | — | — | 12.5 |
| | 6 | — | — | 12.5 |

[a] Performed according to the procedures described in "Principles of Drilling Mud Control," 11th ed., American Petroleum Institute, 1962.
[b] Aqueous suspension of 25 lb/bbl Wyoming bentonite, 5.0 lb/bbl lime, and 2.0 lb/bbl sodium hydroxide.

methylated materials and their derivatives. They have been found to be satisfactory for drilling mud thinners by giving results comparable with those obtained with commercial mud thinners (Table I). Their ability to lower surface tension in aqueous solution indicates that these compounds may also find use as emulsifiers for the *in situ* recovery of the Athabasca bitumen (Table II) (*18*). These materials and other similar derivatives of the asphaltenes, especially those containing functions such as carboxylic or hydroxyl, may also readily exchange cations and could well compete with synthetic zeolites. Other uses of the hydroxyl derivatives and/or the chloro-asphaltenes include high temperature packings and heat transfer media.

Table II.  Surface Tensions of Ozonized and
Ozonized–Sulfonated Bitumen

| Concentration (% w/w) | Surface Tension (dynes/cm) | |
|---|---|---|
| | Ozonized Bitumen | Ozonized–Sulfonated Bitumen |
| 0.1 | 62 | 48 |
| 0.5 | 62 | 41 |
| 1.0 | 52 | 39 |

**Water-Soluble Asphaltenes with Commercial Thinners**

| Viscosity[a] | Plastic Viscosity (centipoise) | Yield Point (lb/100 sq in.) | Gel Strength | |
|---|---|---|---|---|
| | | | 10 sec | 10 min |
| 77 | 7 | 63 | 20 | 32 |
| 17 | 5 | 7 | 2 | 3 |
| 21 | 6 | 9 | 1 | 4 |
| 19 | 7 | 5 | 0 | 4 |
| 22 | 10 | 2 | 1 | 2 |
| 18 | 8 | 2 | 0 | 1 |
| 18 | 8 | 2 | 0 | 1 |
| 13 | 6 | 1 | 0 | 4 |
| 13 | 6 | 1 | 0 | 0 |
| 13 | 6 | 1 | 0 | 0 |
| 14 | 5 | 4 | 3 | 5 |
| 14 | 6 | 2 | 0 | 2 |
| 19 | 7 | 5 | 0 | 7 |
| 12 | 5 | 0 | 0 | 0 |
| 13 | 6 | 1 | 0 | 0 |
| 12 | 5 | 2 | 0 | 2 |

The reactions incorporating nitrogen (up to 14%) and phosphorus (up to 10%) into the asphaltenes are particularly significant at a time when the effects on the environment of many materials containing these elements are receiving considerable attention. Here we have potential slow-release soil conditioners which will only release the nitrogen or phosphorus after considerable weathering or bacteriological action. One may suggest further that the carbonaceous residue remaining after release of the hetero elements may be a benefit to humus-depleted soils such as the grey-wooded and solonetzic soils found in Alberta. It is also feasible that coating a conventional quick-release inorganic fertilizer with a water-soluble or water-dispersible derivative will provide a slower-release fertilizer and an organic humus-like residue. In fact, variations of this theme are multiple. Only further work will tell how practical these projected uses may be, and none should be discounted as long as research continues and the need for better use of petroleum heavy ends remains.

### Acknowledgment

The authors are indebted to Syncrude Canada, Ltd., for the bitumen samples.

*Literature Cited*

1. Speight, J. G., *Fuel* (1970) **49,** 76.
2. *Ibid.* (1971) **50,** 102.
3. *Ibid.* (1973) **52,** 300.
4. Moschopedis, S. E., Speight, J. G., *Fuel* (1971) **50,** 211.
5. *Ibid.,* 34.
6. Rao, C. N. R., "Chemical Applications of Infrared Spectroscopy," p. 305, Academic, New York, 1963.
7. Moschopedis, S. E., Canadian Patent **722,720** (1965).
8. Moschopedis, S. E., U.S. Patent **3,352,902** (1967).
9. Moschopedis, S. E., *Fuel* (1962) **41,** 425.
10. Moschopedis, S. E., Speight, J. G., *Fuel* (1971) **50,** 58.
11. *Ibid.* (1970) **49,** 335.
12. Speight, J. G., *Fuel* (1971) **50,** 175.
13. Moschopedis, S. E., Speight, J. G., *Fuel* (1971) **50,** 332.
14. Speight, J. G., unpublished data.
15. Sawatzky, H., Montgomery, D. S., *Fuel* (1964) **43,** 453.
16. Sawatzky, H., Boyd, M .L., Montgomery, D. S., *J. Inst. Petrol.* (1967) **53,** 162.
17. Speight, J. G., *Am. Chem. Soc., Div. Fuel Petrol. Chem., Preprint* **15,** 57 (1971).
18. Moschopedis, S. E., Speight, J. G., "Abstracts of Papers," 167th Natl. Meet., Am. Chem. Soc., March 1974, Fuel 45.

RECEIVED December 17, 1974.

# 13

# Direct Zinc Chloride Hydrocracking of Sub-bituminous Coal and Regeneration of Spent Melt

CLYDE W. ZIELKE, WILLIAM A. ROSENHOOVER, and EVERETT GORIN

Research Division, Conoco Coal Development Co., Library, Pa. 15129

*Excellent conversion and yields are obtained in the hydrocracking of sub-bituminous coal with a molten zinc chloride catalyst. A vehicle with hydrogen-donor properties is required at relatively mild operating conditions, i.e., 358°C and a hydrogen pressure of 103 atm. No vehicle is necessary at more severe conditions such as 385°C and a hydrogen pressure of 205 atm. Data on the regeneration of the spent catalyst in a continuous fluidized-bed combustion unit show that high recoveries of zinc chloride are obtained if hydrogen chloride gas is recycled to the combustor. The only components of coal ash that are not removed in the process are the alkali metals.*

The use of molten metal halides of the Lewis acid type for the hydrocracking of coal and coal extract was extensively investigated by Consol Research (the predecessor of Conoco Coal Development Co.) during 1964–1967 under a research contract with the Office of Coal Research. A complete description of the work is available in reports to the OCR (*1, 2*). Summary papers dealing with some of this work have also been presented (*3, 4, 5, 6*).

The above work concentrated most of its attention on the use of zinc chloride as the molten halide and on the use of bituminous coal extract as feed to the process. Hydrocracking of the extract (*1*) and regeneration by a fluidized-bed combustion technique of the spent catalyst melt (*2*) from the process were both demonstrated in continuous bench-scale units.

A substantial program was also previously conducted in a batch auto-clave unit on the direct hydrocracking of bituminous coal (1) with zinc chloride melts, but no work was done in either batch or continuous units on regeneration of spent melts from direct hydrocracking of coal.

Other workers also examined direct hydrocracking of coal with molten metal halide catalysts. Several metal halide catalysts, for example, were examined for the direct hydrocracking of bituminous coal (Illinois No. 6) by Wald and Kiovsky (7–13). However, little or no data are available either on the direct hydrocracking of lower rank coals with zinc chloride catalyst or on the regeneration of spent melts from such an operation. A large number of different coals ranging in rank from lignite to bituminous were tested more recently in unpublished work at Consol Research. Almost all coals tested responded well to the zinc chloride hydrocracking technique.

Various procedures for regeneration of spent melts from metal halide hydrocracking were proposed by Kiovsky and Petzny (14, 15) and by Loth and Wald (16). A number of regeneration procedures were also discussed previously in detail (2).

The fluid-bed combustion method (2) has been chosen, however, for process development in the regeneration of spent melts from the hydro-cracking of coal. In this method, from one to two parts by weight of spent melt is generated for each part of coal fed to the hydrocracking process. The carbonaceous residue, sulfur, and ammonia retained in the melt are burned out with air in a fluidized bed of inert solids. The zinc chloride is simultaneously vaporized, the ash separated from the overhead vapors, and the zinc chloride vapor is condensed as pure liquid for return to the process.

The present paper presents batch autoclave data on the direct hydro-cracking of a single sub-bituminous coal from the Powder River basin of southeastern Montana. Comparative data were also obtained with the Pittsburgh Seam bituminous coal that was used in the previous work (1). Data on the regeneration of simulated spent melts from such an operation are also given in a continuous bench-scale, fluidized-bed combustion unit.

### Experimental

**Zinc Chloride Hydrocracking—Batch Autoclave Work.** All tests were made in a 316 stainless steel, 300-ml rocking autoclave. The equip-ment, the product work-up, analytical and calculational procedures used are all identical to those previously described (1). A constant hydrogen partial pressure was used in each run by monitoring it with a palladium–silver alloy probe within the authoclave. The sensitivity of the probe response was increased as compared with prior work by heat treating at

590°C for 4 hr before use. Sub-bituminoues coal from the Colstrip Mine in southeastern Montana was used in this work, and its analysis is given in Table I. The residence time at temperature was 60 min in all the runs reported here.

### Table I. Analysis of Feedstocks

| | | Coals | | Regeneration Feedstock Spent Melt |
|---|---|---|---|---|
| | Colstrip | Cleaned Ireland Mine | Ireland Mine | |
| *Proximate and Ultimate (Wt % MF Basis)* | | | | |
| Volatile matter | 37.3 | 43.4 | 42.5 | — |
| Non-oxidized ash | 9.7 | 6.4 | 9.1 | 1.78 |
| Organic hydrogen | 4.4 | 5.3 | 5.3 | 0.11 |
| Carbon | 68.0 | 76.5 | 74.5 | 5.03 |
| Organic nitrogen | 1.1 | 1.6 | 1.1 | 0.09 |
| Oxygen (by difference) | 16.1 | 8.3 | 7.7 | 0.24 |
| Organic sulfur | 0.5 | 1.9 | 2.3 | 0.04 |
| Pyritic sulfur | 0.2 | 0.9 | 1.7 | — |
| Sulfate sulfur | 0.04 | 0.0 | 0.08 | — |
| $ZnCl_2$ | — | — | — | 85.34 |
| ZnS | — | — | — | 2.92 |
| $NH_4Cl$ | — | — | — | 3.20 |
| $NH_3$ | — | — | — | 1.25 |
| *Ash Composition (Oxidized and $SO_3$-Free Basis)* | | | | |
| $P_2O_5$ | — | — | — | 0.15 |
| $SiO_2$ | 43.9 | 41.0 | 43.5 | 42.2 |
| $Al_2O_3$ | 26.0 | 20.4 | 24.6 | 25.3 |
| $Na_2O$ | 0.25 | 0.6 | 0.5 | 0.47 |
| $K_2O$ | 0.15 | 1.5 | 1.6 | 0.21 |
| CaO | 17.3 | 2.8 | 1.7 | 18.7 |
| MgO | 6.7 | 0.7 | 0.7 | 7.1 |
| $Fe_2O_3$ | 4.7 | 29.6 | 22.8 | 4.6 |
| $TiO_2$ | 1.0 | 1.1 | 1.0 | 1.4 |

**Regeneration by Fluid-Bed Combustion.** PREPARATION OF FEEDSTOCK. During the hydrocracking process, the zinc chloride catalyst becomes contaminated with zinc sulfide, $ZnCl_2 \cdot NH_3$, and $ZnCl_2 \cdot NH_4Cl$ that are formed by the catalyst partially reacting with the sulfur and nitrogen liberated from the feed in the hydrocracking step:

$$ZnCl_2 + H_2S \rightarrow ZnS + 2\ HCl$$

$$ZnCl_2 + xNH_3 \rightarrow ZnCl_2 \cdot xNH_3$$

$$ZnCl_2 \cdot yNH_3 + yHCl \rightarrow ZnCl_2 \cdot yNH_4Cl$$

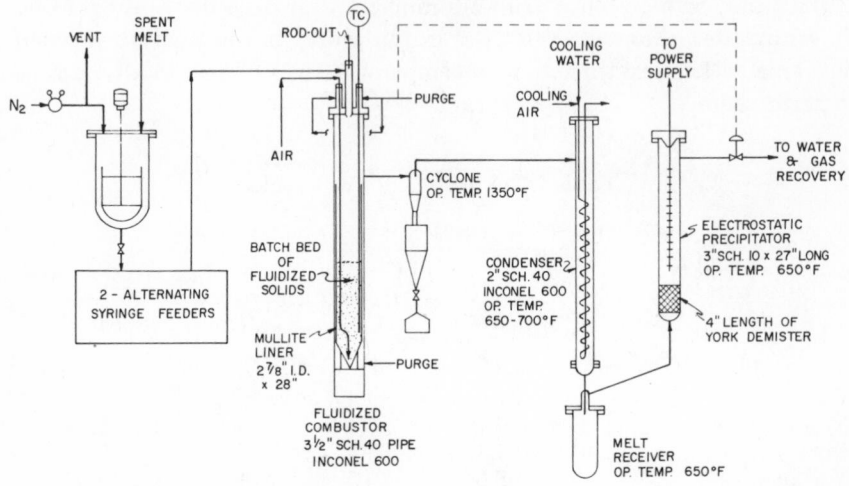

*Figure 1.    Fluidized combustion unit*

The proportions of ZnCl · $y$NH$_3$ and ZnCl$_2$ · $y$NH$_4$Cl depend on the ratio of nitrogen to sulfur in the feed. In addition to these inorganic compounds, the catalyst leaving the hydrocracker also contains residual carbon that cannot be distilled out of the melt. In the case of direct coal hydrocracking, the catalyst also contains the coal ash.

The melt used in this work was prepared from the residue of hydrogen-donor extraction of Colstrip coal with tetralin solvent in such a way as to simulate the composition of an actual spent melt. The extraction was conducted in the continuous bench-scale unit previously described (*17*) at 412°C and 50 min residence time. The residue used was the solvent-free underflow from continuous settling (*17*) of the extractor effluent. The residue was then precarbonized to 675°C in a muffle furnace. The melts were blended to simulate the composition of a spent melt from the direct hydrocracking of the Colstrip coal by blending together in a melt pot zinc chloride, zinc sulfide, and ammonium chloride, ammonia, and the carbonized residue in appropriate proportions. Analysis of the feed melt used in this work is given in Table I.

EQUIPMENT AND PROCEDURE. Figure 1 is a diagram of the continuous, 2⅞-in. id fluidized bed combustion unit that was used. The melt is fed *via* syringe feeders and is dropped from a remote drip tip into a batch bed of fluidized solids that is fluidized by feed air that enters at the apex of the reactor cone. The carbon, nitrogen, and sulfur are burned out in the fluidized bed, and the zinc chloride is vaporized. The gas, zinc chloride vapor, and elutriated solids leaving the reactor pass through the cyclone where the solids are collected. The cyclone underflow solids derive solely from the melt since the sizing of the bed solids is such that there is essentially no elutriation of this material. The solids collected at the cyclone then include coal ash, zinc oxide formed by hydrolysis of zinc chloride, and any unburned carbon or zinc sulfide. The gas then passes to the condenser where zinc chloride is condensed out, then to the electrostatic precipitator to remove zinc chloride fog, and then to sampling and metering. The analytical methods and calculational procedures are substantially the same as those previously described (2).

### Results and Discussion

**Hydrocracking.** A series of experiments was carried out at a relatively low temperature and hydrogen partial pressure of 358°C and 103 atm, respectively. Comparison runs were carried out with Pittsburgh

### Table II. Hydrocracking at Mild Conditions

*Constant Conditions for All Runs*

| | |
|---|---|
| Temperature | 358°C |
| Time at temperature | 60 min |
| Total time above 315°C | 70 min (approx.) |
| $H_2$ partial pressure during hydrocracking | 103 atm. abs. |
| Coal particle size | −100 mesh |

*Variable Conditions and Yields*

| Coal | Colstrip | Ireland Mine | Cleaned Ireland Mine |
|---|---|---|---|
| Feed (g)/MF coal (g) | | | |
| $ZnCl_2$ | 2.5 | 2.5 | 2.0 |
| solvent | 0.5 | 0.5 | 0.0 |
| solvent used | tetralin | tetralin | none |
| | | | |
| Yields (wt % MAF coal) | | | |
| $C_1$–$C_3$ hydrocarbons | 1.7 | 5.1 | 5.2 |
| $CO + CO_2$ (gas) | 5.5 | 0.8 | 0.1 |
| $H_2O$ | 12.3 | 7.2 | 8.7 |
| $C_4 \times 400$°C distillate | 27.5 | 43.8 | 56.4 |
| +400°C MEK soluble | 46.9 | 31.3 | 20.3 |
| +400°C MEK insoluble | 10.5 | 16.5 | 12.8 |
| N,O,S, + H to catalyst | 1.5 | 2.8 | 3.0 |
| total | 105.9 | 107.5 | 106.5 |
| | | | |
| $H_2$ Consumed (wt % MAF coal) | 5.9 | 7.5 | 6.5 |
| | | | |
| Wt % conversion MAF coal | | | |
| to − 400°C products | 42.6 | 52.2 | 66.9 |
| to − 400°C + MEK solubles | 89.5 | 83.5 | 87.2 |

Seam coal from the Ireland Mine used in the previous work. Selected results are given in Table II. The previous work with bituminous coal at these mild conditions was carried out without a hydrocarbon vehicle. These data are reproduced in the last column of Table II.

Attempts to run with the sub-bituminous coal at the above mild conditions without a vehicle were not successful because a high viscosity mix formed which made temperature control very poor and yielded erratic results.

Methylnaphthalene was then used as a vehicle, but even in this case very poor results were obtained, i.e., conversion to methyl ethyl ketone (MEK) solubles and distillate products was very low. Results of this run are not reported since they are obscured by extensive solvent cracking.

The use of tetralin as a hydrogen-donor solvent, however, gave a very good conversion to MEK soluble products as shown in Table II. In this case, some cracking of the solvent also occurred, but the hydrogen consumption and gaseous products are calculated neglecting any contribution made to these quantities by hydrogenation and cracking of the tetralin.

The addition of a hydrogen-donor solvent to the sub-bituminous coal at these mild conditions, appears to assist in melting the coal to permit access of the molten halide catalyst.

The highly fluid Pittsburgh Seam coal, on the other hand, does not require addition of a vehicle as the data of Table II show. As a matter of fact, superior results were obtained in the absence of a vehicle although the difference may arise from the fact that in one case a cleaned coal was used, i.e., the poorer results with the vehicle may reflect some adverse effect of the mineral matter on the hydrocracking process. Also, a somewhat lower catalyst/coal feed ratio was used in the run without a vehicle.

Operation at these mild conditions is of interest where the objective is to produce low sulfur fuel oil in major amounts as a coproduct with gasoline. Previous work showed that 65–80% of the MEK solubles may be recovered from the spent melt by extraction with a fraction of the distillate oil product. The data of Table III, interestingly enough, show that the MEK soluble oil contains less than 0.2 wt % sulfur even when the high sulfur Pittsburgh Seam coal is used as feedstock.

The total liquid yield, i.e., $C_4$ through MEK soluble hydrocarbons at these mild conditions, is almost as high with the sub-bituminous coal

### Table III. Sulfur Content of +400°C MEK Soluble Oil

| Coal | Colstrip | | Ireland Mine | |
|---|---|---|---|---|
| Temperature (°C) | 358 | 385 | 358 | 385 |
| H$_2$ partial pressure (atm) | 103 | 205 | 103 | 205 |
| % Sulfur in MEK solubles | 0.17 | 0.04 | 0.17 | 0.18 |

### Table IV. Hydrocracking at Severe Conditions

*Constant Conditions for All Runs*

| | |
|---|---|
| Time at temperature | 60 min |
| Total time above 315°C | 70 min |
| $ZnCl_2$/MF coal feed ratio | 2.5 |
| $H_2$ partial pressure | 205 atm abs. |
| Coal particle size | −100 mesh |

*Yields and Conversions (Wt % MAF Coal)*

| Coal | Colstrip | Colstrip | Ireland Mine |
|---|---|---|---|
| Temperature (°C) | 385 | 413 | 385 |
| $CH_4$ | 0.6 | 1.3 | 0.4 |
| $C_2H_6 + C_3H_8$ | 4.8 | 8.6 | 8.3 |
| $C_4 \times 200°C$ distillate | 55.6 | 59.0 | 66.4 |
| $200 \times 400°C$ distillate | 3.1 | 1.2 | 2.2 |
| +400°C MEK soluble | 17.1 | 9.2 | 13.9 |
| +400°C MEK insoluble | 5.2 | 8.7 | 6.0 |
| $CO + CO_2 + H_2O$ | 19.9 | 19.3 | 8.6 |
| N,O,S, + H to catalyst | 2.4 | 2.2 | 3.3 |
| total | 108.7 | 109.5 | 109.1 |
| $H_2$ Consumed (wt % MAF coal) | 8.7 | 9.5 | 9.1 |
| Wt % Conversion MAF Coal | | | |
| to − 400°C products | 77.7 | 82.1 | 80.1 |
| to − 400°C + MEK solubles | 94.8 | 91.3 | 94.0 |
| Total − 400°C hydrocarbons | 64.1 | 70.1 | 77.3 |
| Total − 400°C hydrocarbons/$H_2$ consumed | 7.4 | 7.4 | 8.5 |

(74.4 wt % of the moisture- and ash-free (MAF) coal) as with the bituminous coal (75.5 and 76.7 wt % of the MAF coal with and without the use of a vehicle, respectively). The hydrogen consumption, on the other hand, is significantly lower for the sub-bituminous case. This is primarily because of the lower gas and distillate yields. The yield of heavy fuel oil, i.e., MEK soluble oil, is significantly higher in the case of the sub-bituminous coal.

If the temperature and pressure are increased, then a vehicle is no longer necessary even with the sub-bituminous coal. A series of experiments was carried out without a vehicle using a hydrogen partial pressure of 205 atm and at temperatures ranging from 370° to 427°C. The zinc chloride/moisture free (MF) coal feed ratio and the reaction time were held constant at 2.5/1 and 60 min, respectively, in these runs.

The above series of runs showed that two optimum temperatures existed. The first, at about 385°C, gave a maximum conversion to MEK-soluble and lighter products of about 95%. The second, at about 410°C,

### Table V.    General Conditions

| | |
|---|---|
| Pressure (atm abs.) | 1.14 |
| Superficial air velocity (cm/sec) | 30.5 |
| Superficial residence time (sec) | 1.0 |
| Melt feed rate (kg/hr/m²) | 356 |
| % of stoichiometric air | 115 |
| Fluidized bed depth (cm) | 30.5 |
| Type of bed solids | 28 × 48 M silica |

gave a maximum conversion to gasoline expressed as $C_4 \times 200°C$ distillate of about 60 wt % of the MAF coal.

More detailed results of two runs, each near the above two optimum temperatures are given in Table IV. Comparison is also given with results obtained wih Ireland Mine coal at the lower temperature.

The conversion and hydrogen consumptions are very nearly the same for both coals. The yield of hydrogen distillates and particularly gasoline is about 20% higher for the bituminous coal case. The hydrogen consumption per unit of liquid distillate hydrocarbons produced is about 15% higher for the sub-bituminous coal. The above differences are to be anticipated and are a result of the higher hydrogen and lower oxygen contents of the bituminous coal as noted in Table I. The sulfur content of the MEK soluble oil is again very low as noted in Table IV.

**Regeneration via Fluid-Bed Combustion.** Table V gives the conditions of runs presented here—essentially atmospheric pressure, superficial residence time of 1 sec, excess air (115% of stoichiometric), silica bed solids. Temperatures of 983° and 1038°C were investigated. No

### Table VI.    Distribution of Carbon, Sulfur, and Nitrogen in the Products

| Run number | 3 | 4 |
|---|---|---|
| Temperature (°C) | 983 | 1038 |
| % of stoichiometric air | 115 | 115 |
| Distribution of C (%) | | |
|     burned to $CO_2$ (+ loss) | 92.7 | 92.9 |
|     burned to CO | 5.9 | 4.2 |
|     in cyclone solids | 0.4 | 0.5 |
|     in melt | 1.0 | 2.4 |
| Distribution of S (%) | | |
|     burned to $SO_2$ (+ loss) | 91.3 | 96.4 |
|     in bed | 1.8 | 0.4 |
|     in cyclone solids | 3.8 | 1.4 |
|     in melt | 3.1 | 1.8 |
| Distribution of N (%) | | |
|     $NH_3$ burned to $N_2 + H_2O$ (+ loss) | 89.3 | 100.0 |
|     N in melt | 10.7 | 0.0 |

operability problems such as ash agglomeration or defluidization of the bed were encountered at these conditions.

Table VI gives the distribution of carbon, sulfur, and nitrogen in the products for runs at 983°C and 1038°C. One-hundred fifteen percent of stoichiometric air was used in both runs. The feed gas was pure air. It is apparent that the carbon, nitrogen, and sulfur impurities were almost completely burned out, *i.e.*, 89% or more burn-out was achieved. The effluent melt contained less than 3% of the carbon and sulfur in the feed melt and 11% or less of the nitrogen.

Table VII shows the distribution of total coal ash and the coal ash components, silica and alumina, in the products of two runs at 983°C.

**Table VII.    Distribution of Total Coal Ash, Silicon, and Aluminum Among the Products**

| Run number | 3 | 11 |
|---|---|---|
| Temperature (°C) | 983 | 983 |
| Feed gas composition (mol %) | | |
| air | 100 | 94.5 |
| anhydrous HCl | — | 5.5 |
| Distribution of total ash (%) | | |
| bed solids (+ loss) | 19 | 34 |
| cyclone solids | 70 | 59 |
| melt | 11 | 7 |
| Distribution of Si (%) | | |
| bed solids (+ loss) | 29 | 28 |
| cyclone solids | 67 | 68 |
| melt | 4 | 4 |
| Distribution of Al (%) | | |
| bed solids (+ loss) | } 94 | 31 |
| cyclone solids | | 64 |
| melt | 6 | 5 |

In run 3, the feed gas was 100% air while in run 11, the feed gas was 94.5% air and 5.5% anhydrous hydrogen chloride. The hydrogen chloride was added to prevent hydrolysis of the zinc chloride and is discussed later. Eighty-nine percent or more of the total coal ash and silica and alumina were removed from the melt in the regeneration process. The ashes were either trapped in the bed or collected at the cyclone.

Table VIII shows the distribution of the ash components, iron, calcium, and magnesium in the products of the same two runs as in Table VII. Again, it is apparent that most of these ash components were removed from the melt and that they were either trapped in the combustor bed solids or collected at the cyclone.

**Table VIII. Distribution of Iron, Calcium, and Magnesium among the Products**

| | 3 | 11 |
|---|---|---|
| Run number | 3 | 11 |
| Temperature (°C) | 983 | 983 |
| Feed gas composition (mol %) | | |
|   air | 100 | 94.5 |
|   anhydrous HCl | — | 5.5 |
| Distribution of Fe (%) | | |
|   bed solids (+ loss) | } 73 | 0 |
|   cyclone solids | | 61 |
|   melt | 27 | 39 |
| Distribution of Ca (%) | | |
|   bed solids (+ loss) | 29 | 53 |
|   cyclone solids | 48 | 42 |
|   melt | 23 | 5 |
| Distribution of Mg (%) | | |
|   bed solids (+ loss) | 44 | 57 |
|   cyclone solids | 43 | 41 |
|   melt | 13 | 2 |

The distribution of sodium and potassium is not given in the tables. Essentially all of the sodium and potassium in the feed appears in the effluent melt. This is likely caused by the potassium and sodium being converted into the chlorides which are sufficiently volatile to be vaporized along with the zinc chloride. This chlorine is considered irrecoverable. Hence, this method of regeneration is chiefly useful with coals that have relatively low alkali concentrations.

Table IX shows the distribution of zinc and chlorine among the products in runs 3 and 11. The feed gas in run 3 was pure air. In this run, about 10% of the feed zinc was found in the bed solids and cyclone solids as zinc oxide while 90% was in the melt as zinc chloride. The large amount of zinc oxide was formed by hydrolysis of zinc chloride in the combustor, according to:

$$ZnCl_2 + H_2O = ZnO + 2 HCl$$

Additional loss of zinc oxide by reaction with the silica to form zinc silicate is also a possibility. Because of the hydrolysis, a large amount of the chlorine in the feed melt is found in the gas as hydrogen chloride. For the process to be economically feasible, zinc chloride would have to be reformed from zinc oxide and hydrogen chloride products.

To preclude such a step, the hydrogen chloride can be recycled with the feed air to prevent hydrolysis. Run 11 was made to test this concept where the feed gas contained 5.5% anhydrous hydrogen chloride. The chlorine in the feed hydrogen chloride amounted to 14.8% of the chlorine in the zinc chloride feed. It is apparent that hydrolysis was almost completely suppressed and that the zinc in the feed melt was almost totally converted to and recovered as zinc chloride.

The zinc lost to the bed solids with hydrogen chloride recycle was reduced to 0.4 wt % of the zinc fed. This shows that hydrogen chloride not only prevented hydrolysis of zinc chloride but also prevented any significant quantity of zinc silicate from forming.

Before starting this work, it was feared that considerable chlorine would be lost as calcium chloride by interaction of the calcium in the coal ash with the zinc chloride, but it appears that essentially no chlorine is lost in this manner. In both runs 3 and 11, the bed solids contain substantially no chlorine whereas they contained a large percentage of the calcium that was fed. Since calcium chloride is molten but nonvolatile at combustion temperature, it would be expected that any calcium chloride would be retained in the bed solids. Since none was, it is concluded that no calcium chloride was formed. It also appears that no magnesium chloride was formed.

Table X shows some pertinent reactions in the regeneration system. We determined the equilibrium constant for Reaction 3 and obtained $P_{ZnCl_2}$ of 57 torr whereas $P_{ZnCl_2}$ in the combustor was about 120 torr. Hence, it might be expected that some calcium chloride would be formed. It is believed that the reason we did not get any is because of reactions such as Reactions 4 and 5 whose equilibria are probably far to the right.

### Table IX.   Distribution of Zinc and Chlorine in the Products

| | | |
|---|---|---|
| Run number | 3 | 11 |
| Temperature (°C) | 983 | 983 |
| Feed gas composition (mol %) | | |
| air | 100 | 94.5 |
| anhydrous HCl | — | 5.5 |
| Cl distribution (% of Cl in feed melt) | | |
| bed solids | .02 | .01 |
| cyclone solids | .44 | .30 |
| gas | 9.10 | 14.50 |
| melt (+ loss) | 90.40 | 100.00 |
| Total | 100.00 | 114.80 |
| Zn distribution (% of Zn in feed melt) | | |
| bed solids | 5.5 | 0.40 |
| cyclone solids | 4.4 | 0.74 |
| melt (+ loss) | 90.1 | 98.90 |

Table X.  Some Pertinent

*Reaction*

(1) $ZnCl_{2(v)} + H_2O_{(g)} = ZnO_{(s)} + 2\ HCl_{(g)}$

(2) $ZnCl_{2(v)} + FeO_{(s)} = FeCl_{2(v)} + ZnO_{(s)}$

(3) $CaCl_{2(l)} + ZnO_{(s)} = CaO_{(s)} + ZnCl_{2(v)}$

(4) $CaCl_{2(l)} + SiO_{2(s)} + ZnO_{(s)} = CaSiO_{3(s)} + ZnCl_{2(v)}$

(5) $CaCl_{2(l)} + SiO_{2(s)} + H_2O_{(g)} = CaSiO_{3(s)} + 2\ HCl_{(g)}$

Kuxman and Oder (*18*) reported recovering zinc as pure zinc chloride vapor from impure ores by Reaction 4 carried out at about 900°C.

It appears that some ferrous chloride is formed but not in the amount that would be expected if equilibrium had been established in Reaction 2. Reactions analogous to Reactions 4 and 5 with ferrous chloride substituted for calcium chloride may be the reason for this. In any case, the equilibrium constant indicates that the amount of ferrous chloride that can be formed is limited to one mol per nine mols of zinc chloride. We have unpublished data that indicates that ferrous chloride does not affect catalyst activity.

We believe, based on the results just presented and on other results, that fluid-bed combustion provides a workable process for regenerating zinc chloride from direct hydrogenation of western sub-bituminous coals. Other work not presented here indicates that the process can also be applied successfully to melts from direct hydrogenation of eastern bituminous coals. The process is restricted, however, to coals having relatively low sodium and potassium contents so that economically prohibitive amounts of chlorine are not lost to these alkali metals. Lignites are the major type of coal that would be ruled out by the above restriction.

*Literature Cited*

1. Consolidation Coal Co., R & D Report No. 39, Office of Coal Research, U. S. Dept. of the Interior. Interim Report No. 2, "Research on Zinc Chloride Catalyst for Converting Coal to Gasoline—Phase I—Hydrocracking of Coal and Extract with Zinc Chloride," Vol. III, Book 1, March 1968.
2. Consolidation Coal Co., R & D Report No. 39, Office of Coal Research, U. S. Dept. of the Interior. Interim Report No. 2, "Pre-Pilot Plant Research on the CSF Process—Phase I—Regeneration of Zinc Chloride Catalyst," Vol. III, Book 2, March 1968.
3. Zielke, C. W., Struck, R. T., Evans, J. M., Costanza, C. P., Gorin, E., "Process Design Develop.," *Ind. Eng. Chem.* (1966) **5**, 151.
4. *Ibid.*, 158.

**Reactions**

$$K \qquad\qquad K \text{ at } 983°C$$

$$K_1 = \frac{(P_{HCl})^2}{(P_{ZnCl_2})(P_{H_2O})} \qquad .28$$

$$K_2 = \frac{P_{FeCl_2}}{P_{ZnCl_2}} \qquad\qquad .11$$

$$K_3 = P_{ZnCl_2} \qquad\qquad 57 \text{ torr}$$

5. Struck, R. T., Clark, W. E., Dudt, P. J., Rosenhoover, W. A., Zielke, C. W., Gorin, E., *Ind. Eng. Chem.* (1969) **8**, 546.
6. Zielke, C. W., Struck, R. T., Gorin, E., *Ind. Eng. Chem.* (1969) **8**, 552.
7. Wald, M. M., British Patent **1,310,283** (March 14, 1973).
8. Kiovsky, T. E., U.S. Patent **3,764,515** (October 9, 1973).
9. Wald, M. M., U.S. Patent **3,542,665** (1970).
10. Kiovsky, T. E., Wald, M. M., U.S. Patent **3,668,109** (1972).
11. Wald, M. M., U.S. Patent **3,824,178** (1974).
12. Kiovsky, T. E., Wald, M. M., U.S. Patent **3,725,239** (1973).
13. Kiovsky, T. E., Wald, M. M., U.S. Patent **3,663,452** (1972).
14. Kiovsky, T. E., Petzny, W. J., U.S. Patent **3,685,692** (1972).
15. Kiovsky, T. E., Petzny, W. J., U.S. Patent **3,657,108** (1972).
16. Loth, R. H., Wald, M. M., U.S. Patent **3,790,469** (1974).
17. Gorin, E., Lebowitz, H. E., Rice, C. H., Struck, R. T., *Proc. World Pet. Congr., 8th* (1971) **PD10**, 5.
18. Kuxmann, V., Odor, F., Z. *Erzbergbau Metallhuttenwes.* (1966) **19**, 388.

RECEIVED December 17, 1974.

# 14

# Differences Among Ozokerites

ROBERT F. MARSCHNER and J. C. WINTERS

Research and Development Department, Amoco Oil Co.,
Naperville, Ill. 60540

*Ozokerites are mainly mixtures of n-alkanes that occasionally accompany deposits of petroleum, coal, or lignite. Gas chromatograms of 10 ozokerites from Galicia, Utah, and Russia showed systematic differences in composition. In all samples from Galicia, n-alkanes near $C_{29}$ were most abundant. This same maximum probably occurs in hatchettites and Uinta Basin petroleum. Some Utah samples also peaked near $C_{29}$, but in others the maximum occurred near $C_{46}$, with a secondary peak around $C_{31}$. The Russian ozokerites, n-alkanes near either $C_{31}$ or $C_{46}$ are the more abundant. Multiple favored abundance ranges suggest that a sequence of proccesses were involved in the formation of ozokerites.*

$D$eposits of mineral wax occasionally associated with petroleum, coal, lignite, and other native organic substances are known by a profusion of names. If found near oil fields, they are called earth waxes or ozokerites; if found near coal beds, they are usually called hatchettites; if found near lignite, they are sometimes called scheererite; and specimens found in various environments have been called aragotite, evenkite, fichtelite, hartite, idrialite, posepnyite, valaite, etc. One waxy mineral, kabaite, found in traces in meteorites, seems not to be a member of the same family and may result from hydrogenation of carbon monoxide (*1*). Otherwise, variations within most of these minerals are about as large as the differences between them, and the names have consequently not been used rigorously. A generic term ozokerites could suffice for all, and it is so used here.

Ozokerites are widely distributed, having been found in perhaps a hundred localities around the world. Some properties of ozokerites from a selection of major deposits are given in Table I. Many of these deposits are in the mountainous regions of eastern Europe. The densest occurrence is in Galicia along the northeast flank of the Eastern Carpathian Mountains, especially near the town of Borislav, now in the USSR but, since the turn of the century, also in Poland and Austria–Hungary. It occurs with a variety of inorganic minerals and varies in color from yellow to brown and in melting point from about 40° to 80°C.

n-Alkanes are clearly the characteristic components of ozokerites. Densities and refractive indices are too low for anything but paraffinic hydrocarbons, and the high melting points and low solubilities preclude much of anything other than the normal ones. Unquestionably some

Table I.    Characteristics of Ozokerites from Major Deposits

| Location | Area (Country) | Mineral Association | Color | M.P. (°C) | Refer- ence |
|---|---|---|---|---|---|
| *Associated with Petroleum (Ozokerite)* | | | | | |
| Borislav | Galicia (USSR) | sandstone | yellow | 60–80 | 2 |
| Starunya | Ukraine (USSR) | gypsum | grayish yellow | 77.5 | 3 |
| Shor-su | Fergana (USSR) | limestone | dark | 59–65 | 4 |
| Uinta Basin | Utah (USA) | oil shale | yellow- brown | 60–79 | 5 |
| Slanik | Moldavia (Romania) | shale, coal | — | 93 | 6 |
| *Associated with Coal (Hatchettite)* | | | | | |
| Loch Fyne | Scotland (UK) | — | yellow | 49 | 6 |
| Pencoed | South Wales (UK) | shales | green | 38–65 | 7 |
| Porte de France | Grenoble (France) | limestone | yellowish | 38–29 | 8 |
| *Associated with Lignite (Scheererite)* | | | | | |
| St. Gallen | (Switzerland) | — | reddish | 44–46 | 6 |

branched and cyclic structures are also present (7, 9), and the branched structures include monomethylalkanes and the same group of $C_{15}$–$C_{20}$ isoprenoids that occurs in petroleum (4).

The technology of ozokerites followed a meteoric pattern (10). From first production around 1860 until World War I, use for candles, crayons, electrical insulation, and leather dressings expanded steadily. Such elementary refining as was developed consisted of little more than melting the raw wax in hot water to separate inorganic matter and treating it with concentrated sulfuric acid to produce ceresin, which was lighter in color and higher in price. Uses proliferated between the Wars. At first, ozokerite was widely used as a substitute and adulterant for costly beeswax and plant waxes. Later, cheaper waxes from the expanding petroleum industry were used to extend and replace ozokerite. By World War II, most commercial sources of ozokerite were depleted or even exhausted. Meanwhile, however, vacuum distillation for manufacture of lubricating oils had become widespread in refineries, and what are now marketed as ozokerites and ceresins are actually mixtures of motor-oil waxes from vacuum distillates with microcrystalline waxes from vacuum residues.

Although the origin of ozokerites is not known, they may have been derived from petroleum. Nature may have precipitated ozokerites in a manner parallel to the way that man has precipitated sucker-rod wax in the oil fields or tank-bottom waxes in refineries. The main processes in both instances are removal of light fractions that act as solvent, lowered temperature, and enough time for the wax crystals to consolidate from the viscous medium. Such a mechanism would be difficult to extend to coal or lignite, however, and provenance from unusual plant matter has been proposed as an alternative explanation. Microorganisms also might well have had an influence (11).

Ozokerites in any case provided a logical material for extending previous work on the n-alkanes in crude oils (12). There the abundance of n-alkanes was found to be a significant parameter in the geochemistry.

**Table II.    Descriptions of**

| Source Serial No. (Date) | Original Donor |
|---|---|
| Baryslav, Poland[a] E18150 (1937) | Industrial and Agricultural Museum of Warsaw |
| Boryslaw, Galacia[a] M4963 (1894) | Ward's Natural Science Estab. |
| Galacia, Austria[a] E5015 (1894) | Standard Oil Co. |
| Kyune, Utah[a] E13536 (1910) | private collection |
| Fort Worth, Utah[a] E13067 (1905) | private collection |
| Russia[a] E4979 (1894) | World's Columbian Exposition |
| Carbon County, Utah[b] | Geology Museum |
| Emery County, Utah[b] | Geology Museum |
| H-23 (unknown)[b] | private collection |
| Soldier Summit, Utah[c] | University of Utah |

[a] From Field Museum of Natural History through Bertrand G. Woodland.
[b] From Colorado School of Mines through A. Sherrill Hougton.

In the present study, gas chromatography was again used, with programmed-temperature operation as before, but with longer times at high temperature.

### Experimental

Native ozokerites are not as available as they once were, but pieces do exist in museums and collections. Six samples obtained from the Field Museum of Natural History in Chicago through Bertram G. Woodland served as the nucleus for the present study. Properties of these and supplementary samples are listed in Table II. Especially valuable in view of the clouded history of ozokerites was a description attached to the sample originally from the Industrial and Agricultural Museum in Warsaw: "Earth wax in unrefined state directly after excavation from the mine. From the Association for Earth Wax and Rock Oil Industry at Boryslaw."

Gas chromatograms were obtained on a twin Hewlett-Packard model 5750 Research Chromatograph. The dual columns were 6-m lengths of copper tubing 3 mm in diameter, packed with 3% OV-1 on Chromosorb G-HP (methylsilicone on calcined diatomaceous earth). Samples were introduced as about 10% solutions in carbon bisulfide. Detection was by hydrogen-flame ionization, the non-sample contribution from the idle column being subtracted from the total contribution of the active column to provide a sample chromatograph corrected for extraneous ionization.

### Ozokerites Examined

| Other Description | Melting Point (°C) | Streak on Paper | Approx. % n-Alkanes |
|---|---|---|---|
| *See* text, cast block | 61.2 | orange-brown | 82 |
| sample No. 110 | 63.5 | orange-yellow | 85 |
| commercial | 62.2 | brown-orange | 93 |
| native paraffin | 60.0 | brown | 89 |
| — | 73.6 | light brown | 95 |
| cast cylinder | 83.5 | reddish brown | 70 |
| Soldier Summit | — | light brown | 85 |
| — | — | light brown | 79 |
| probably Utah | 84.0 | orange-brown | 93 |
| Wasatch County, mine sample | 84.6 | dark brown | 95 |

*From Reino E. Kallio, University of Illinois, Urbana, Ill.

The temperature was proprammed from 100°C at 10°C per min to 400°C, where it was held as long as meaningful ionization data were recorded.

These severe conditions pressed the equipment to the limits of operability. Several trials were usually necessary before the complete chromatogram could be accurately defined, and all samples were run two to five times to confirm the presence or absence of elusive characteristics. Columns had to be replaced several times, adjustments to the procedure were frequent, and compromises with consistency and uniformity had to be made continually.

A typical composite chromatogram of three runs on the ozokerite from Russia is shown in Figure 1. Time in minutes from injection of sample is given at the bottom of the figure, and the temperature of the oven in which the columns are heated is given at the top. A partial chro-

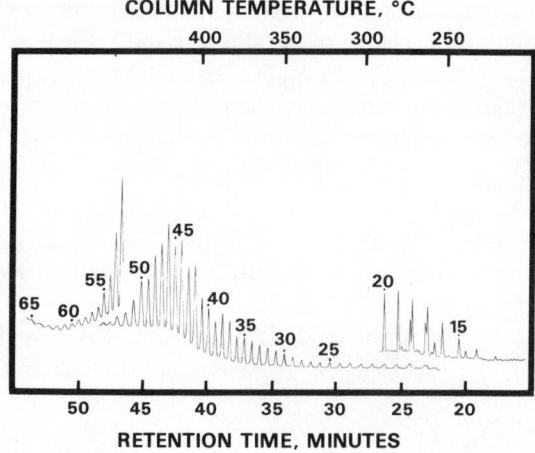

*Figure 1.   Composite chromatogram*

matogram at the right defines the front ends in greater detail, and the one at the left extends the composition 10 carbon numbers higher than the more conventional chromatogram in the center of the figure. The upper-limit temperature of 400°C was reached near $C_{40}$ with this sample. At this point, only about one third of the amount of sample had emerged, although two thirds of the 50 n-alkanes present were already accounted for.

Because n-alkanes are the predominant constituents of ozokerites, determination of carbon number alone identifies most components. Two adjacent doublets near the beginning of some curves located the two pairs, n-heptadecane-pristane and n-octadecane-phytane, which served as unambiguous counters. For samples that started at higher carbon number, longer n-alkanes ($C_{22}$, $C_{28}$, or $C_{32}$) were added to provide internal standards. In addition, the chromatogram of a heavy paraffinic petroleum containing $n$-alkanes from 10 to 50 carbon atoms was obtained frequently as an external standard.

Amounts of individual $n$-alkanes were determined by measuring the areas under the successive peaks and above the baseline that represents an unresolved background of other components. Although the background level rises continually with column temperature, the amount relative to $n$-alkanes increases more from about $C_{30}$–$C_{40}$ than at either lower or higher carbon numbers. Estimates of the $n$-alkane content of all samples are included in Table II.

### n-*Alkane Abundances in Ozokerites*

Compositions were compared through logarithmic distribution plots of the amounts of successive $n$-alkanes. Typical distribution curves show pronounced maxima in certain ranges rather than a gradual diminution with carbon number that might have been expected from previous experience with crude oils (*12*). In many crudes, a maximum occurred at $C_6$, and the $n$-alkanes extinguished around $C_{35}$. The maximum at $C_6$ could have resulted from previous losses of lighter constituents to natural gas or natural gasoline, rather than representing a particularly favored number. The extinction around $C_{35}$ could have resulted from previous separation of longer $n$-alkanes underground or in tankage, rather than no such material having been present at all.

In certain ranges of some crude oils, odd-numbered $n$-alkanes predominate appreciably over the even-numbered homologs. Especially striking are odd predominances from $C_{11}$ to $C_{19}$ that accompany declines in abundance from $C_{16}$ to $C_{18}$ and from $C_{18}$ to $C_{20}$ (*12*). Odd–even predominances were also observed in ozokerites, but not as prominently and at higher carbon numbers; they will be presented and discussed in detail elsewhere.

Galician ozokerites have $n$-alkane distribution curves that almost coincide, as shown by Figure 2. Most abundant are $n$-alkanes near $C_{29}$, although all three samples have a weak shoulder indicating a somewhat favored range around $C_{17}$. The Ward's and Warsaw samples are shown with plateaus in the $C_{50}$'s but the compounds creating them are probably not $n$-alkanes. Peaks in the chromatograms in the $C_{40}$'s of Ward's sample particularly are doublets if well resolved, and those that appear to correspond to the $n$-alkanes gradually disappear by $C_{50}$. The second peaks might suggest adulteration, but a natural contaminant seems more probable. Its amount is barely 1%, and it can be found only if the analysis is carried to a high enough carbon number. The curve for the Standard Oil sample has been corrected for the spurious second peaks to better represent the true $n$-alkane abundances.

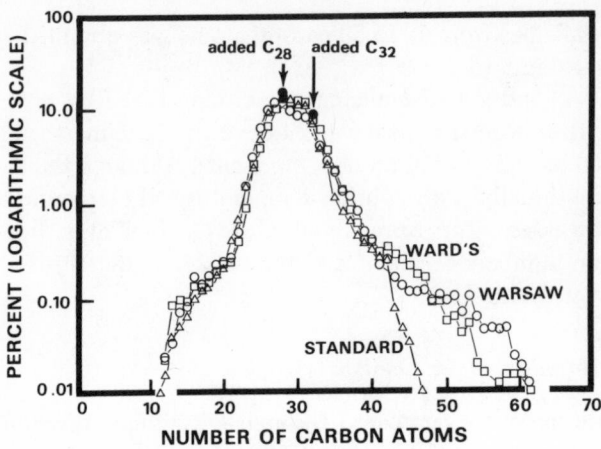

*Figure 2.  Galician ozokerites*

Utah ozokerites have a variety of n-alkane distribution curves, as shown in the three panels of Figure 3. The three ozokerites in the top panel most nearly resemble those from Galicia: n-alkanes near $C_{29}$ are most abundant, and favored ranges are suggested at both lower and higher carbon number. The peaks do not exactly coincide; it is lowest for Kyune, which has a shoulder at lower carbon number, and is highest for H-23, which has the most prominent plateau at higher carbon number. Unlike those of the Galician ozokerites, the gas chromatogram peaks in the $C_{40}$'s are not doubled, and the higher constituents are presumably n-alkanes.

Fort Worth ozokerite in the middle panel of Figure 3 is displaced to the right of the previous curves, the maximum is broader and asymmetrical, and the lesser peaks are more pronounced. The major peak falls at $C_{33}$–$C_{37}$, or about four carbon atoms higher than the Galician ozokerites. It has the appearance of an unresolved doublet, with one peak just above $C_{30}$ and the other just above $C_{40}$. The lower peak falls at $C_{17}$–$C_{19}$, no more than two carbons higher than the Galician, whereas the higher peak falls at $C_{59}$, which is six or eight carbons higher than either the Galician or the other Utah ozokerites.

Finally, in the bottom panel of Figure 3 are two more Utah ozokerites with four distinct abundance regions. The maximum occurs at $C_{43}$–$C_{45}$ and the next largest peak at $C_{31}$–$C_{33}$, just opposite of Galician ozokerites and those from Utah in the top panel. The lowest favored region is not well defined, especially for Soldier Summit, and the identification with the $C_{17}$–$C_{19}$ region of the previous curve is uncertain. Indeed, the Wasatch sample may have more than one favored region in this range.

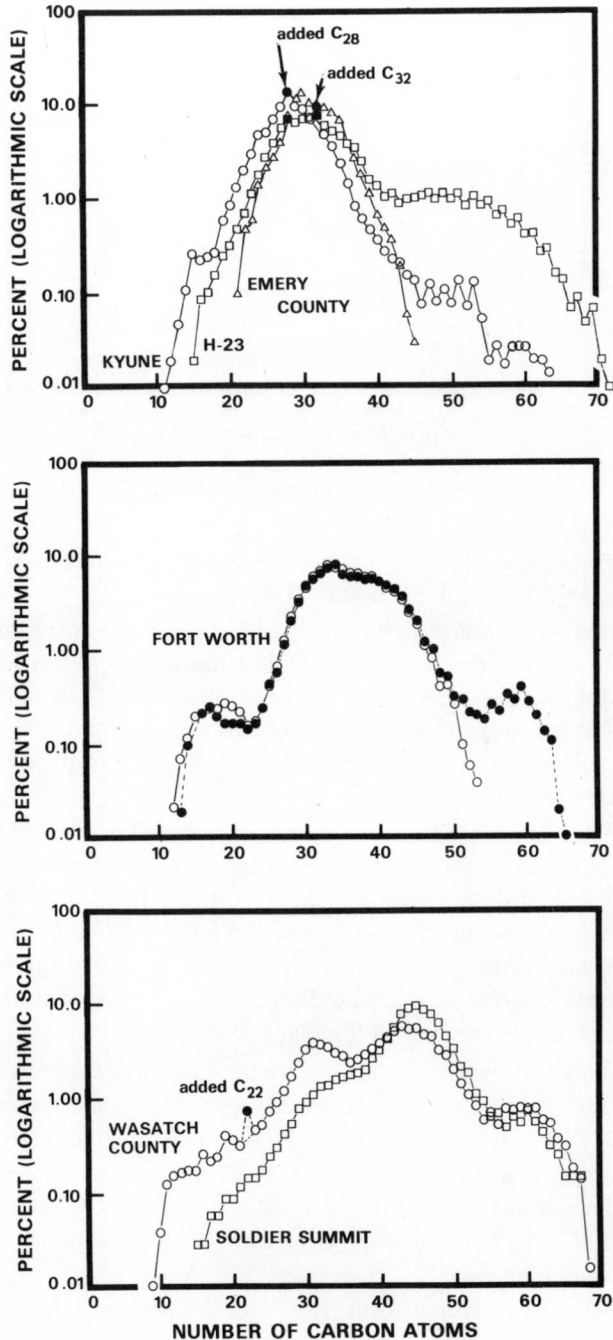

*Figure 3.   Utah ozokerites*

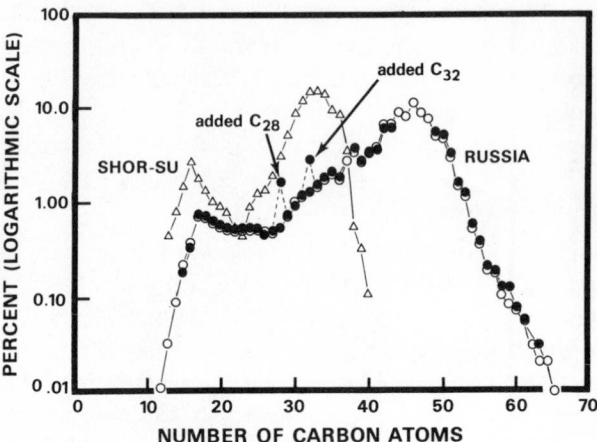

*Figure 4.    Russian ozokerites*

The highest favored region, on the other hand, is well defined at $C_{59}$ or $C_{60}$ and is the same for both samples.

The simpler samples in the top panel came from locations south of the complex ones in the bottom panel, but the source of the Fort Worth sample in the middle panel is not known. Two other Utah samples were indistinguishable from those illustrated. Any or all of them may have contained still longer *n*-alkanes beyond the reach of gas chromatography, but not likely in large amounts.

The Russian ozokerite is compared with the Shor-su ozokerite from southern Fergana, USSR, from the literature (4) in Figure 4. They are different, but again not unrelated. Shor-su is the simpler, with only two constituents, the smaller at $C_{16}$–$C_{18}$ and the larger at $C_{31}$–$C_{34}$. The other appears to have three constituents, represented by a maximum at $C_{46}$, a long plateau at $C_{18}$–$C_{27}$, and a shoulder near $C_{35}$. These lesser abundances may be the same as the two in Shor-su.

Distribution curves for other ozokerites that have come to our attention are shown in Figure 5. Hatchettites from the South Wales coalfield have been thoroughly surveyed (7). A single maximum in *n*-alkane abundance occurs at $C_{24}$–$C_{29}$, with 12 of 18 samples actually peaking at $C_{27}$, but the chromatograms did not extend beyond $C_{35}$, and longer *n*-alkanes could have been missed. Before weathering, another hatchettite showed a shoulder near $C_{16}$ that suggests a lesser abundance at lower carbon number. Ein Humar ozokerite from the eastern side of the Dead Sea in Jordan (13), shows an abundance maximum at $C_{38}$, together with the suggestion of a lower abundance anomaly.

Patterns seen in the Russian ozokerites of Figure 4 and the miscellaneous ozokerites of Figure 5 were previously visible in the Utah ozokerites of Figure 3. The most abundant *n*-alkanes may fall either near $C_{29}$, as in Galician ozokerites, or near $C_{46}$, as in the Soldier Summit type, and implications based on samples from Galicia alone could be misleading.

### Causes of Maxima in Abundance

A single maximum in abundance is deceptively easy to explain. A diminution with increase in carbon number might be expected from the thermodynamics of whatever chain-forming process was originally involved, a diminution with decrease in carbon number might be expected from such separation processes as volatility or solubility that might have occurred subsequently, and the resultant of the two processes would be an intermediate maximum, whether the material under investigation was an ozokerite or a crude oil.

One difficulty with this ready explanation is that the maxima in ozokerite and petroleum are not the same. Three of eight crude oils peaked at $C_{11}$ and three more at $C_7$ or below (*12*). Two other crude oils peaked higher, as shown in Figure 6 (State Line at $C_{19}$ and Uinta Basin at $C_{27}$), but only Uinta Basin has a maximum at a carbon number nearly as high as those in ozokerites. Inspection of Figure 6, however, suggests the possibility of coincidence; Uinta Basin has a distinct plateau near $C_9$ and an indistinct one around $C_{19}$, whereas the broad peak for State Line could result from overlapping of three abundance maxima at $C_9$, $C_{19}$, and $C_{29}$.

Closer examination of *n*-alkane abundances among ozokerites support this possibility. All three Galician samples, for example, show an abun-

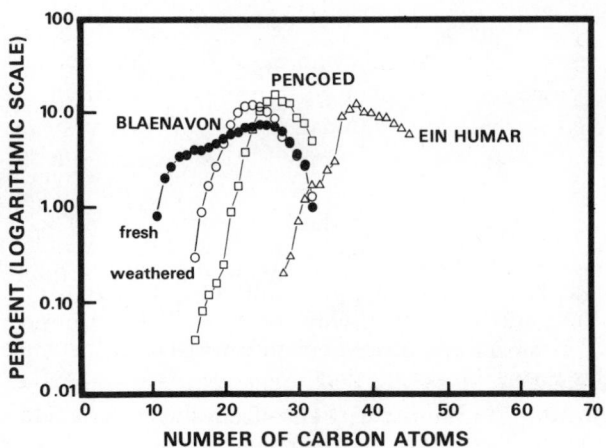

*Figure 5.   Other ozokerites*

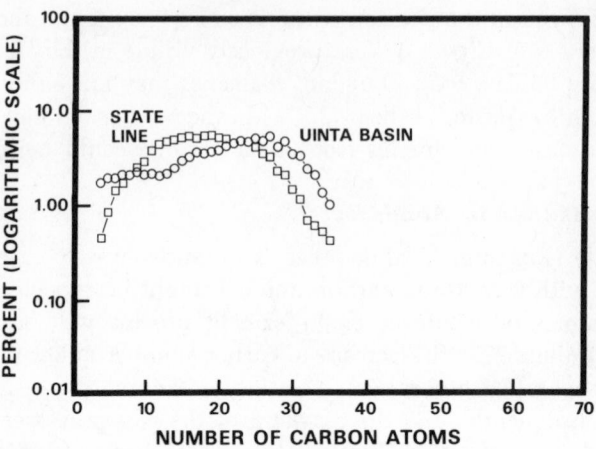

*Figure 6.   Crude oils*

dance shoulder around $C_{18}$ that could correspond to the maximum in
State Line and the indistinct shoulder in Uinta Basin. Russian ozokerite
and Pencoed hatchettite are similar. The lowest abundance anomalies
for Utah ozokerites occur only slightly lower, as also does that for un-
weathered Blaenavon hatchettite.

Multiple maxima such as occur with Russian and some of the Utah
ozokerites are more difficult to explain. Nearly all ozokerites show a
pronounced abundance peak near $C_{31}$, but several show a clear secondary
peak around $C_{46}$, and a few show even a third one in the $C_{50}$s. Variation
also remains to be accounted for. The prime maxima for hatchettites
occur four or more carbon atoms lower, and the secondary peak for Ein
Humar occurs perhaps eight carbon atoms lower.

The presence of more than one favored abundance range suggests that
several processes were involved in ozokerite formation. If the processes occurred in sequence, abundances representing vestiges of earliest
processes as well as n-alkanes formed latest might both be seen. To
account for the distribution of n-alkanes actually found in petroleums and
ozokerites would require at least the following processes.

1. Vaporization. Maxima in n-alkane abundance ranging from $C_5$
to $C_{11}$ in most crude oils probably result from the vaporization of normally gaseous constituents, primarily through reduction in pressure. The
position of the maximum should depend primarily on the gas-to-oil ratio,
greater ratios giving higher maxima.

2. Biosynthesis. The favored $C_{15}$–$C_{19}$ range may represent vestigal
odd-numbered n-alkanes produced by decarboxylation of the even-numbered 16–20 carbon fatty acids produced by plants. Presumably the
mechanism for the biosynthesis is the same for ozokerite as for pertroleum,

but any differences in the positions of the maxima would result mainly from differences in the source plants.

3. Maturation. The low level of odd–even predominance categorizes ozokerites as having undergone maturation in a manner similar to equilibrium crude oils. The light gases involved in vaporization (1.) would also be produced in the maturation process.

4. Evaporation. Maxima in the approximate range $C_{20}$–$C_{30}$ are partly the result of evaporation of lower homologs during long access to the atmosphere. This process is inherent in that of "inspissation," through which deposits of bitumen are conventionally accounted for, and has been implicated in hatchettite formation (7).

5. Melting. Pronounced maxima in the $C_{27}$–$C_{33}$ range may result in part from the loss of smaller *n*-alkanes by a fractional melting process analogous to "sweating" in wax manufacture. Positions of the maxima would be affected by overlap with evaporation (4.).

6. Solubility. Also involved in concentration of *n*-alkanes in the $C_{27}$–$C_{33}$ range may be the immobility of higher *n*-alkanes by reason of insolubility in mixed hydrocarbons underground. Immobility may be assisted by adsorptive forces involving minerals or such organic materials as kerogen.

7. Selective synthesis. Abundant *n*-alkanes in the $C_{38}$–$C_{50}$ range must obviously have been formed by selective synthesis at some stage of the genesis—early, by plants; during diagenesis, perhaps accompanying maturation; or subsequently, by some unfamiliar polymerization process. Perhaps the traces of nonnormal structures in Galician ozokerites provide a clue to the mechanism involved.

8. Inertness. The final favored range of *n*-alkane abundance at $C_{51}$–$C_{59}$ may represent vestiges of an infinite series that survived by reason of inertness through, say, insolubility. On occasion, *n*-alkanes above $C_{80}$ were observed, and homologs as high as $C_{100}$ may be present in minute amounts.

Not all these processes need necessarily have been involved in formation of every petroleum or even every ozokerite, but both seem to have required a considerable combination of circumstances, with ozokerite the more complicated of the two. The variations in its composition, as well as its scarcity relative to petroleum, coal, and lignite, were inevitable results.

### Acknowledgments

We thank Arie Nissenbaum for making the chromatogram for Ein Humar ozokerite available in advance of publication.

### Literature Cited

1. Anders, Edward, Hayatsu, Ryoichi, Studier, Martin H., "Organic Compounds in Meteorites," *Science* (1973) **182**, 781.

2. Redwood, Boverton, "Galician Petroleum and Ozokerite Industries," *J. Soc. Chem. Ind.* (1892) **11**, 93.
3. Kwiecinska, Barbara, Ratajezak, T., "Ozokerite from Starunya," *Bull. Acad. Pol. Sci.* (1969) **17**, 155, *Chem. Abs.* (1970) **73**, 132948.
4. Kovjazin, V. E., Hala, S., "Presence of $C_{15}$-$C_{20}$ Isoprenoid Alkanes in Ozokerite," *Collect Czech. Chem. Commun.* (1973) **38**, 2938.
5. Robinson, Heath M., "Ozokerite in Central Utah," *U.S. Geol. Surv. Bull.* (1916) **641A**.
6. Abraham, Herbert, "Asphalts and Allied Substances," Vol. I, Chapter7, Van Nostrand, 1960.
7. Firth, J. N. M., Eglinton, G., "Hatchettite from the South Wales Coalfield," *Adv. Org. Geochem.* (1971) 613.
8. Louis, M., Bienner, F., "A Specific Paraffinicity Index: Hatchettite," *Rev. Inst. Fr. Pet.* (1954) **9**, 149.
9. Kastner, Dietrich, Moos, Josef, Schultze, Georg R., "Composition of Crude Polish Ozokerite," *Erdoel Kohle* (1959) **12**, 77.
10. Ivanovsky, Leo, "Ozokerite and Related Substances," 3 vols., Hartleben, Leipzig, 1934.
11. Rozanova, E. D., Shturm, L. D., "Changes in Ozokerite Composition under the Action of Microorganisms," *Mikrobiologiya* (1966) **35**, 138.
12. Martin, Ronald L., Winters, John C., Williams, Jack A., "Distribution of *n*-Paraffins in Crude Oils and Implications to Origin of Petroleum," *Nature* (1963) **199**, 110.
13. Nissenbaum, A., Aizenshtat, Z., "Geochemical Studies on Ozokerite from the Dead Sea Area," *Chem. Geol.* (1975) **16** (2), 121.

RECEIVED December 16, 1974.

# INDEX

*The text of this book is set in 10 point Caledonia with two points of leading. The chapter numerals are set in 30 point Garamond; the chapter titles are set in 18 point Garamond Bold.*

*The book is printed offset on Text White Opaque 50-pound. The cover is Joanna Book Binding blue linen.*

*Jacket design by Norman Favin. Editing and production by Virginia Orr.*

*The book was composed by the Mills-Frizell-Evans Co. and by Service Composition Co., Baltimore, Md., printed and bound by The Maple Press Co., York, Pa.*